Encyclopedia of
Mathematics and Society

Mathematics in Culture and Society

Encyclopedia of Mathematics and Society

Mathematics in Culture and Society

Editors
Sarah J. Greenwald
and
Jill E. Thomley
Applachian State University

SALEM PRESS
A Division of EBSCO Publishing
Ipswich, Massachusetts

Cover Photo: © Daniel Ramsbott/dpa/Corbis

Copyright © 2013, by Salem Press, A Division of EBSCO Publishing, Inc.
All rights reserved. No part of this work may be used or reproduced in any manner whatsoever or transmitted in any form or by any means, electronic or mechanical, including photocopy, recording, or any information storage and retrieval system, without written permission from the copyright owner. For permissions requests, contact proprietarypublishing@ebscohost.com.

ISBN 978-1-4298-3751-4

Printed in the United States of America

Contents

List of Contributors	ix
Acrostics, Word Squares, and Crosswords	1
Algebra in Society	3
Archery	9
Ballet	10
Ballroom Dancing	11
Baseball	13
Basketball	15
Basketry	17
Betting and Fairness	19
Billiards	21
Birthday Problem	22
Board Games	24
Calculus in Society	26
Cheerleading	31
Climbing	33
Cocktail Party Problem	34
Composing	35
Connections in Society	37
Cooking	42
Crochet and Knitting	45
Dice Games	46
Extreme Sports	48
Fantasy Sports Leagues	50

Fireworks	52
Fishing	54
Football	55
Geometry in Society	57
Geometry of Music	63
Golden Ratio	65
Gymnastics	67
Harmonics	69
Hockey	71
Knots	73
Lotteries	74
Magic	75
Marriage	78
Martial Arts	80
Mathematical Puzzles	82
Mathematics and Religion	84
Mathematics Genealogy Project	89
Measurement in Society	91
Musical Theater	96
Numbers and God	98
Origami	101
Painting	102
Percussion Instruments	104
Popular Music	106
Predicting Divorce	108
Puzzles	109
Pythagorean and Fibonacci Tuning	113
Quilting	115
Racquet Games	117
Rankings	118
Representations in Society	119
Scales	125
Sculpture	127
Six Degrees of Kevin Bacon	128
Skating, Figure	130
Skydiving	131

Soccer	132
Sport Handicapping	134
Square Dancing	135
Step and Tap Dancing	136
String Instruments	137
Sudoku	138
Swimming	139
Symmetry	140
Textiles	143
Tic-Tac-Toe	144
Volleyball	145
Wind Instruments	147

Resource Guide   **149**

List of Contributors

Stephen Abbott
Middlebury College
Hyungryul Baik
Cornell University
Eric Barth
Kalamazoo College
John Beam
University of Wisconsin, Oshkosh
Bonnie Ellen Blustein
West Los Angeles College
Casey Borch
University of Alabama at Birmingham
John N. A. Brown
Independent Scholar
Diana Cheng
Middle Tennessee State University
Dogan Comez
North Dakota State University
Richard De Veaux
Williams College
Maria Droujkova
Natural Math
Lee Anne Flagg
University of Alabama at Birmingham
Catherine C. Galley
Independent Scholar
Deborah L. Gochenaur
Shippensburg University
Lidia Gonzalez
City University of New York
Michael K. Green
State University of New York, Oneonta
Sarah J. Greenwald
Appalachian State University

Juan B. Gutierrez
University of Miami
Holly Hirst
Appalachian State University
Linda Hutchison
University of Wyoming
Jerry Johnson
Western Washington University
Michael "Cap" Khoury
University of Michigan, Ann Arbor
Michael Klucznik
St. Bonaventure University
Matt Kretchmar
Denison University
Bill Kte'pi
Independent Scholar
Michele LeBlanc
California Lutheran University
Stephen Lee
Mathematics in Education and Industry
Chad T. Lower
Pennsylvania College of Technology
Philip McCartney
Northern Kentucky University
Elizabeth A. McMillan-McCartney
Northern Kentucky University
Liliana Monteiro
Affiliation TK
Ashwin Mudigonda
Universal Robotics Inc.
Samuel Obara
Texas State University
Eoin O'Connell
Deakin University

Serkan Ozel
Bogazici University

Zeynep Ebrar Yetkiner Ozel
Fatih University

David C. Royster
University of Kentucky

Carl R. Seaquist
Texas Tech University

Lawrence H. Shirley
Towson University

Jorge Nuno Silva
University of California, Berkeley

Henrik Sorensen
Aarhus University

Kristi L. Stringer
University of Alabama at Birmingham

Jill E. Thomley
Appalachian State University

Marcella Bush Trevino
Independent Scholar

K. G. Valente
Colgate University

Matthew West
University of Alabama at Birmingham

Sharon Whitton
Hofstra University

Elizabeth L. Wilmer
Oberlin College

Acrostics, Word Squares, and Crosswords

Category: Games, Sport, and Recreation.
Fields of Study: Geometry; Number and Operations; Problem Solving.
Summary: Mathematics and symmetry come into play in creating and solving word puzzles.

Acrostics, word squares, and crossword puzzles are the most common forms of word puzzles in English. Acrostics and word squares are over 2000 years old and call for the solver to discover words hidden either covertly (acrostics) or overtly (word squares). The crossword puzzle premiered in 1913 and is similar to a word square expanded onto a larger grid, with gaps. Word puzzles have been used as mnemonics, ciphers, literary devices, educational exercises, and as simple games. Their construction, especially in the case of crossword puzzles, is informed by geometry; their solution can be pursued through probability theory. In a sense, the construction and solving of word puzzles provide pleasures very similar to those of doing mathematics.

Historic Examples

The earliest examples of acrostics are in the Old Testament of the Bible. The Lamentations of Jeremiah and 12 Psalms are arranged so that the first letters of each verse spell out the Hebrew alphabet.

In Greece in 400 B.C.E., Dionysius forged a Sophoclean text titled *Parthenopaeus* with the intention of mocking his rival, Heraclides. Having declared the author to be Sophocles, Heraclides was referred to in one of the several acrostics that Dionysius had included, which read, "Heraclides is ignorant of letters."

In more contemporary times, novelist Vladimir Nabokov enjoyed chess problems, and one can find acrostics, number puzzles, cryptic references, and puns in several of his novels and stories. The last paragraph of his 1951 short story "The Vane Sisters," for example, can be read both as the narrator's confusion and acrostically (taking the first letter of each word) as a message from the dead sisters.

Acrostics are often found in poetry because of its greater flexibility in syntax and phrasing. Former U.S. President George Washington is known to have constructed at least one acrostic when he was 15—a love poem for a girl about whom nothing is known other than her name, Frances.

Another good example of an acrostic poem is to be found at the end of Lewis Carroll's 1871 book *Alice Through the Looking Glass*; each letter of the name Alice Pleasance Liddell begins a new line in the poem about childhood innocence.

Word Squares

If the first acrostics appeared in the Old Testament, word squares were not far behind. One of the most well known is a Latin word square from about 2000 years ago:

S	A	T	O	R
A	R	E	P	O
T	E	N	E	T
O	P	E	R	A
R	O	T	A	S

This word square is called a 5-by-5 symmetric word square because there are five words that can be read either down or across. The words "TENET," "OPERA," and "ROTAS" will be familiar to speakers of languages descended from Latin. SATOR is a Latin word for planter or creator. AREPO is a contentious word;

it can be assumed that it was at some time used in Latin. This particular word square is unique in another way—SATOR reversed is ROTAS, AREPO is OPERA reversed, and TENET is palindromic (reads the same forward and backward).

Below is an example of an ordinary symmetrical 4-by-4 word square using English words

```
B A S E
A W A Y
S A L E
E Y E S
```

Many 5-by-5 and 6-by-6 squares exist in English. There are even a few 9-by-9 word squares, though many of the constituent words are extremely unfamiliar.

Those with an interest in algebra will notice that symmetry in word squares is equivalent to symmetry in matrices. If one transposes—swaps the rows and columns—a symmetrical word square, the resulting word square is the same as the original. A non-symmetrical word square does not have this property. A 4-by-4 double word square, like the one below, is not symmetrical. It is a double word square because it contains twice the number of words of a 4-by-4 symmetrical square, that is, eight:

```
D A R T
O B O E
C L A M
K E M P
```

Crosswords

Word squares can be entertaining in themselves. However, simply by expanding a word square onto a larger grid and using gaps to section long words into shorter ones, one can create a puzzle of an altogether different kind. By doing so, puzzle creator Arthur Wynne turned the largely esoteric practice of crafting word squares into a puzzle for the masses—the crossword.

The first published crossword appeared in December 1913 in the newspaper *New York World*. Wynne wrote definitions for each of the words he had used to complete a diamond-shaped grid, and it was up to the solvers of the newspaper's puzzle page to fill in the blanks.

Wynne's grid was almost fully "checked," which means that most letters were part of two words—a white square is "unchecked" when it is part of only one word. In U.S. crosswords, it remains the norm to have very heavily—if not fully—checked grids. For other crossword types, particularly cryptic crosswords, grids may be only 50% to 60% checked. Having a fully checked grid means that it is possible to complete the crossword by entering only the across (or down) words. As the number of unchecked squares increases, however, the ability to build on one's correct answers decreases. Most crosswords have a 15-by-15 grid and twofold rotational symmetry (they look the same after 180 degrees of rotation), but differences in the number of checked squares can produce as many as 80 words or as few as 30.

PROVERB, a computer program designed to solve crosswords, relies on the heavily checked nature of American-style grids. Computer scientist Michael Littman and others report that PROVERB averaged more than 95% correct answers in less than 15 minutes per puzzle on a sample of 370 puzzles. This result is better than average human solvers but not better than the best. If nothing else, the complexity of the PROVERB program serves to highlight the vast computing power humans naturally possess.

Instinctively, many people may not be aware that the five most frequently used letters in the English language are E, T, A, O, and I. Crosswords setters (and PROVERB), on the other hand, are acutely aware of this and aim to use letters in their longer words that will be easy to intersect with the shorter ones. It is therefore worth bearing in mind that, for example, "Erie" and "Taoist" will appear in crosswords much more often than "jazz" and "Quixote." Incidentally, the five least frequently used letters are K, J, X, Z, and Q.

Estimates suggest that fewer than 100 people construct crossword puzzles for a living in the United States. Mathematician Byron Walden has been called "one of the best" by a *New York Times* crossword editor. For some, he may be most well known for writing the puzzle that was used in the championship round of the American Crossword Puzzle Tournament, later featured in the film *Wordplay*. He has also analyzed and given talks on symmetry and patterns associated with conventional crossword construction, with the aim of helping people become more skilled puzzle solvers.

Mathematician Kiran Kedlaya is also a well-known puzzle solver and creator. He believes that the brain processes required for computer science, mathematics, music, and crossword puzzles are similar, and he pursues all of these activities professionally and recreationally. One puzzle he created was published on the well-known *New York Times* crossword page, and he regularly contributes mathematics puzzles to competitions like the USA Mathematical Olympiad. He has been quoted as saying, "It's important to tell kids who are interested in math as a career that there are many venues to do it, not just in the academic area within math departments."

Further Reading
Balfour, Sandy. *Pretty Girl in Crimson Rose (8)*. Sirlingshire, UK: Palimpset Book Production, 2003.
Littman, M., et al. "A Probabilistic Approach to Solving Crossword Puzzles." *Artificial Intelligence* 134 (2002).
MacNutt, Derrick Somerset. *Ximenes on the Art of the Crossword*. London: Methuen & Co., 1966.

Eoin O'Connell

Algebra in Society

Category: School and Society.
Fields of Study: Algebra; Connections.
Summary: Algebra provides tools for orderly thinking and problem solving, applicable across a spectrum of pursuits.

Among the many discussions in his 1961 book *The Realm of Algebra*, science fiction author and biochemist Isaac Asimov described the real-life uses of algebra; explored the role it played in the discoveries of scientists and mathematicians such as Galileo Galilei and Sir Isaac Newton; and suggested the idea that "the real importance of algebra, and of mathematics in general, is not that it has enabled man to solve this problem or that, but that it has given man a new outlook on the universe." This notion underlies many of the perspectives on algebra in the twenty-first century.

Knowledge of algebra is seen as important not only for scientific research and the workplace but also for teaching general logical thinking and for making decisions that are important to personal well-being and society as a whole. For example, some functional relationships among people's day-to-day activities that may affect personal decisions include the relationship between how much food a person eats and weight; the amount of exercise and weight loss; and calculations for loans, interest, and other financial matters.

Some would say that the ramifications of these relationships and a lack of understanding of them mathematically are found in the housing crisis of the early twenty-first century and the increase in obesity. Algebra is reported as being a challenging subject for some people.

Many consider algebra to be a major gateway into higher mathematics in both high school and college, and it is thus critical to careers in engineering, science, mathematics, and other disciplines that require advanced mathematics training. Performance of primary and secondary students on algebra tests is one common comparison measure used to evaluate the relative standing of countries with regard to education. Professional organizations like the National Council of Teachers of Mathematics (NCTM) continue to examine the role of algebra in society and make recommendations. Some of the numerous careers that have been cited as requiring algebra include architecture, banking, carpentry, dentistry, civil engineering, nursing, pharmacy, and plumbing.

How Is Algebra Useful?
In 2003, the RAND Corporation's Mathematics Study Panel underscored the key role of algebra in education by choosing it as one of the panel's main areas of focus, explaining their decision in part by saying, "Algebra is foundational in all areas of mathematics because it provides the tools (i.e., the language and structure) for representing and analyzing quantitative relationships, for modeling situations, for solving problems, and for stating and proving generalizations." In algebra, there are general laws or algebraic models that can be used to represent a given scenario.

Algebra is sometimes noted as a type of language that provides answers to all cases at all times and models the relationships between quantities, reducing the need for repeated or inefficient computation. For example, in order to determine the savings in an interest-bearing account after a given period of time, one could compute the savings each month or year by multiplying by the interest rate. However, this computation is cumbersome after many compounding cycles. Instead, the algebraic formula

$$A = P(1+r)^t$$

can be applied directly, where P is the initial investment, r is the interest rate per period, t is the number of periods, and A is the amount of money in the bank after t periods. People may want to know if it is profitable to leave money in a bank subjected to the stated formula. On the other hand, people may want to determine the present and future value of the money they have invested because of the effect of inflation. In other instances, such as taking a car or home loan, similar algebraic laws exist. These laws help people know how much money, for instance, they may save if they pay off their loan earlier than the due date. In the eleventh century, scholar, poet, and mathematician Omar Khayyam explained the following:

> . . . Algebra is a scientific art. The objects with which it deals are absolute numbers and measurable quantities which, though themselves unknown, are related to "things" which are known, whereby the determination of the unknown quantities is possible. . . . What one searches for in the algebraic art are the relations which lead from the known to the unknown. . . . The perfection of this art consists in knowledge of the scientific method by which one determines numerical and geometric unknowns.

Early History

Algebra definitions and applications have evolved over time, though many aspects of algebraic thinking and methods that are taught in twenty-first-century schools can be traced back to antiquity. The Babylonians and Egyptians used algebraic techniques to solve problems directly related to the everyday needs of society, such as dividing land and keeping financial records. One such example from Babylonian mathematics is an alterna-

Clifford Ho, 2010 Asian American Engineer of the Year in Sandia National Laboratory's solar power heliostat field. Studying algebra is critical to pursuing careers in engineering, science, and mathematics. (Sandia National Laboratories, Randy Montoya)

tive method for solving cubic equations of the form $x^3 + x^2 = b$, via tabulated numerical values of squares and cubes. The Babylonians were able to solve this polynomial by using the table that gave the values of $x^3 + x^2$ or $x^2(x+1)$. They constructed the table to solve: $x^2(x+1) = 1;30$ in sexigesimal notation. The "periods" below are used to represent multiplication.

x	$x^3 + x^2$
1	1.2 = 2
2	4.3 = 12
3	9.4 = 36
4	16.5 = 80
5	25.6 = 150
6	36.7 = 252
7	49.8 = 392
8	64.9 = 576
9	81.10 = 810
10	100.11 = 1100
.	
30	900.31 = 27900

The algorithm used by the Babylonians to find the roots of cubic equations is different from the modern approach, although it can be explained using modern language.

For example, in modern notation, in solving the equation $x^3 + 2x^2 - 3136 = 0$ set $x = 2y$. Then the equation can be rewritten as the following:

$$(2y)^3 + 2(2y)^2 - 3136 = 0$$
$$8y^3 + 8y^2 - 3136 = 0$$
$$y^3 + y^2 = 392.$$

From the table, $y = 7$. Since $x = 2y$, then $x = 14$.

Topics that are viewed as algebra in contemporary mathematics were often numerical or geometric in nature. The Pythagorean theorem, named for Pythagoras of Samos, can be expressed in terms of the algebraic equation that relates the sum of the sides surrounding a right angle in a triangle squared to the square of the hypotenuse. However, historically, there is evidence that the Babylonians explored numerical versions of the theorem, while the Greeks examined the areas of the geometric squares that sat on the edges of the triangle.

The Pythagorean theorem can be found in twenty-first-century algebra classrooms, and it is useful in setting right angles in constructions and in measuring distance in flat objects. Symbolic notation for algebra was developed in India and became popular in Europe in the seventeenth and eighteenth centuries. Historical methods reflect the unique construction of understanding, indicative of the localized culture at that time. Algebraic methods have also been found in some ancient Chinese works.

Greeks, Hindus, Arabs, Persians, and Europeans all contributed to the development of algebra. The term itself comes from the Arabic word *al-jabr*, which has several translations including "the science of equations." The word appears in the title of the early algebra text written by Muhammad Ibn Musa Al-Khowarizmi in the ninth century.

Applied Algebra

For a long time, one major emphasis in algebra was solving polynomial equations, but in the eighteenth century, algebra went through a transformation that broadened the field to include study of other mathematical structures. Around that time, textbooks defined algebra in many different ways. According to mathematician Colin Maclaurin, "Algebra is a general method of computation by certain signs and symbols which have been contrived for this purpose, and found convenient.

It is called an universal arithmetic, and proceeds by operations and rules similar to those in common arithmetic, founded upon the same principles." Leonhard Euler defined algebra as: "The science which teaches how to determine unknown quantities by means of those that are known." As the concept of variables was further developed, many physical properties, including time, mass, density, pressure, temperature, charge, and energy, were expressed algebraically.

For instance, Albert Einstein's equation relates energy to mass times the speed of light squared. In the twenty-first century, defining algebra commonly requires a broader approach. First, one could say that early or elementary algebra is essentially the study of equations and methods for solving them; and second, that modern or abstract algebra is the study of various mathematical structures. High school algebra textbooks typically contain a breadth of topics, such as polynomials and systems of linear equations. These are important in modeling many relationships in society. For example, parabolas represent the paths of ball or bullet trajectories, and systems of linear equations and matrices give rise to digital images. At the college level, students continue their study of algebraic equations in virtually every mathematics and statistics class. Students in a broad range of majors, including the sciences and mathematics, may further their understanding of systems of linear equations and their applications in a linear algebra class.

Mathematics majors in modern or abstract algebra study topics like groups, rings, and fields, and graduate students further explore these and other algebraic structures. These concepts have been useful in chemistry, computer science, cryptography, crystallography, electric circuits, genetics, and physics. Algebra is a core area from the middle grades and high school to undergraduate and graduate mathematics. Research fields include the connections of algebra with other subdisciplines, like algebraic geometry, algebraic topology, or algebraic number theory, and the abstract structures and notions in pure algebra have been applied in many contexts. Some algebraists work for the National Security Agency and others work as professors.

In general, mathematicians and scientists often algebraically derive laws for a given scenario or relationship from patterns. For example, consider a triangle number pattern. It is fairly simple to find the next number recursively but finding larger values such as the 1000th triangular number without a general rule can be more challenging. (See Figure 1 and Table 1.)

Algebra can be used to generalize the preceding case and derive that

$$1 + 2 + 3 + \cdots + n = \frac{n(n+1)}{2}$$

so the general law will be

$$a_n = \frac{n(n+1)}{2}.$$

Hermann Weyl noted "The constructs of the mathematical mind are at the same time free and necessary. The individual mathematician feels free to define his notions and set up his axioms as he pleases. But the question is will he get his fellow mathematician interested in the constructs of his imagination. We cannot help the feeling that certain mathematical structures which have evolved through the combined efforts of the mathematical community bear the stamp of a necessity not affected by the accidents of their historical birth. Everybody who looks at the spectacle of modern algebra will be struck by this complementarity of freedom and necessity."

Many algebraic equations are used in everyday life to meet societal needs. For example, the area of a rectangle is given by the length times the width. There are algebraic equations like finding the area of a square or circle, and also finding volume, which are used in applications like home decorating, cooking, landscaping, and construction. Building houses and fences, determining amounts of material needed for a project, and completing everyday chores use algebra to make work accurate and efficient. Economists use algebraic laws to project business profits or losses and to advise investors and other decision makers. In other instances such as taking a car or home loan similar algebraic laws help people know how much money, for instance, they may save if they pay off their loan earlier than the due date.

Many formulas are easy to use and can easily be entered in a hand calculator or computer to generate the required result. Such formulas have been adapted

Figure 1.

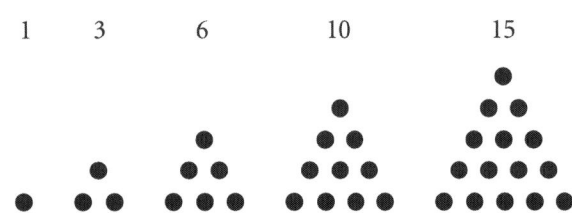

Table 1.

1st term	2nd term	3rd term	4th term	5th term	nth term
1 = 1	3 = 1 + 2	6 = 1 + 2 + 3	10 = 1 + 2 + 3 + 4	15 = 1 + 2 + 3 + 4 + 5	= 1 + 2 + 3 + $\cdots$ + n

to Web-based applets and software like spreadsheets to track financial records, making them widely accessible and often easy to use.

Mathematician Roger Cooke explained, "Algebra provided more than just a compact notation for writing down relations among variables. Its rules made it possible to manipulate those laws on paper and derive some of them from others. For example, a consequence of Kepler's third law is that the ratio . . . of the square of a planet's period to the cube of its distance from the sun is the same for all planets . . . Kepler's third law and Newton's law of gravitation are equivalent statements, given certain basic facts of mechanics." Kepler's laws were named for Johannes Kepler and Newton's for Sir Isaac Newton. The ability to express algebraic relationships using variables and rates of change in calculus increased the applicability of equations in a wide variety of contexts.

The U.S. Bureau of Labor Statistics highlights the importance of coursework in algebra for numerous careers, including brickmasons, blockmasons, and stonemasons; carpenters; computer control programmers and operators; construction and building inspectors; engineers and engineering technicians; line installers and repairers; machine setters, operators and tenders in metal and plastic; machinists; opticians; physical therapist assistants; power plant operators, distributors and dispatchers; sheet metal workers; surveyors, cartographers, photogrammetrists, and surveying and mapping technicians; radiation therapists; tool and die makers; and veterinarians.

Algebra's Role as a Gateway

Some would argue that in the United States, mathematics achievement has not met the same standards of excellence as in other developed countries, and that, as a result, students may not be prepared to enter college. Some historians trace the growing need for mathematics education to the turn of the twentieth century or the Industrial Revolution, when there were debates about the appropriate level of mathematics for high school education. Historically, popular opinion was often against algebra as a subject of widespread study in secondary schools, since many did not see clear connections between algebra and real-world needs. Mathematics educator W. D. Reeve cited one newspaper editorial as an example of such an attitude in a 1936 National Council of Teachers of Mathematics report, *The Place of Mathematics in Modern Education*, saying the following:

> Quite frankly, I see no use for algebra except for the few who will follow engineering and technical lines. . . . I cannot see that algebra contributes one iota to a young person's health or one grain of inspiration to his spirit. . . . I can see no use for it in the home as an aid to a parent, a citizen, a producer, or a consumer.

The same report noted deficits in algebra skills even among graduate students and relatively high failure rates for algebra students in some high schools, such as in New York City, which were used by some as additional arguments against algebra's broad inclusion in the high school curriculum. With regard to who should and should not study algebra, Reeve countered: ". . . no one, I think, has the wisdom to decide who will profit most by its study or predict who the future Newtons and .s are to be."

Mathematician and philosopher Alfred Whitehead stated the following:

> Quadratic equations are part of algebra, and algebra is the intellectual instrument which has been created for rendering clear the quantitative aspects of the world. There is no getting out of it. Through and through the world is infected with quantity. To talk sense, is to talk in quantities. It is no use saying that the nation is large. . . . How large? It is no use saying that radium is scarce. . . . How scarce? You cannot evade quantity. You may fly to poetry and to music, and quantity and number will face you in your rhythms and your octaves. . . . This question of the degeneration of algebra into gibberish, both in word and in fact, affords a pathetic instance of the uselessness of reforming educational schedules without a clear conception of the attributes which you wish to evoke in the living minds of the children. . . . First, you must make up your mind as to those quantitative aspects of the world which are simple enough to be introduced into general education; then a schedule of algebra should be framed which will about find its exemplification in these applications.

Other newspapers like the *Columbus Dispatch* supported broad high school mathematics education during that time period, asserting that schools should provide the "mathematical key" to the "gateways of a larger life."

Algebra eventually became commonplace in high schools and some middle schools, with basic algebraic concepts often introduced even in the primary grades, and yet questions about how to teach algebra continued. Algebra is usually a prerequisite for all higher mathematics courses in both high school and college, and, in some cases, it is required for high school graduation. Students will not advance in many majors or career paths unless they pass algebra, and the result is that some students change majors or abandon education altogether. Students requiring remediation courses at the college level are fairly common.

The result is that in the twenty-first century, algebra is still viewed by many as a major gatekeeper to educational and career advancement, and learning algebra has been promoted as a civil rights issue for every U.S. citizen, though many of the same arguments from past decades continue to be debated. In the latter twentieth century, algebra education became a renewed topic of discussion from local school districts all the way to the White House. The RAND Corporation's panel further explained its decision to focus on algebra by saying, "Without proficiency in algebra, students cannot access a full range of educational and career options, and this curtailment of opportunities often falls most directly on groups that are already disadvantaged."

At the same time, naysayers continue to publish counterpoints regarding algebra's lack of utility. One 2006 *Washington Post* article about a student named Gabriella, who purportedly dropped out of high school after failing her algebra course many times, asserted that writing teaches logical reasoning more effectively than algebra and stated that many students will "never need to know algebra" in the real world, since most mathematics can now be done by computer or calculator. It concluded that having an algebra requirement for high school graduation is potentially more detrimental than helpful because it may spur students to drop out who otherwise might have graduated.

This article spurred many further discussions, and it appeared to reflect the author's own difficult experiences with algebra, a phenomenon that has been reported by many educational researchers and that drives further curricular revisions. Authors of algebra textbooks and self-help books have explored different ways to help students connect to algebra. For instance, actress Danica McKellar has written algebra readiness and algebra books that include stories and characters in order to express equations and solutions in contextual situations. Some educators incorporate mnemonics, songs, or other memory techniques such as First, Outside, Inside, Last (FOIL) in order to teach the multiplication of two binomials. Other authors highlight real-life applications, historical connections, or solutions using technology.

Many national reports have indicated that education in the United States is in a critical period, and some would say particularly in mathematics and science. Educators and politicians have proposed changes to the mathematics education curriculum to prepare U.S. students. The number of students entering college and requiring courses that enable them to be effective in the workplace is rising. Further, engineering and other technical fields that were once seen as elite or remote are increasingly a part of daily life, including computing, electronics, business, and architecture. Technology is changing every day, which has changed society, including mathematics. As a result, there is an increased need for people who can adapt to the changes and continue being effective in society. In this context, there has been a movement to reform algebra education so that it can be more readily accessed by everyone. The "algebra for all" movement has been a central point within the reform initiatives. National standards such as those published by the NCTM have stressed the need to make algebra more accessible to students, and they often outline both the content to be covered and instruction expectations. Some research has shown that students who take algebra by eighth or ninth grade are more likely to pursue higher mathematics, though this cannot be interpreted as a cause-and-effect relationship.

Further Reading

Cohen, Richard. "What Is the Value of Algebra?" *Washington Post* (February 16, 2006). http://www.washingtonpost.com/wp-dyn/content/blog/2006/02/15/BL2006021501989.html.

Cooke, Roger. *Classical Algebra: Its Nature, Origins, and Uses*. Hoboken, NJ: Wiley-Interscience, 2008.

Edwards, Edgar. *Algebra for Everyone*. Reston, VA: National Council of Teachers of Mathematics, 1990.
Eves, Howard. *An Introduction to the History of Mathematics*. Philadelphia: Saunders College, 1990.
The Futures Channel. "Algebra in the Real World." http://www.thefutureschannel.com/algebra_real_world.php.
Gallian, Joseph. *Contemporary Abstract Algebra*. 7th ed. Belmont, CA: Brooks Cole, 2009.
McKellar, Danica. *Hot X: Algebra Exposed: Word Problems, Polynomials, Quadratic Equations and More*. New York: Hudson Street Press, 2010.
Rappaport, Josh. *Algebra Survival Guide: A Conversational Guide for the Thoroughly Befuddled*. Santa Fe, NM: Singing Turtle Press, 1999.
United States Bureau of Labor Statistics. "Occupational Outlook Handbook." http://www.bls.gov/oco.
Zaccaro, Edward. *Real World Algebra*. Bellevue, IA: Hickory Grove Press, 2001.

Samuel Obara

Archery

Category: Games, Sport, and Recreation.
Fields of Study: Algebra; Geometry; Measurement.
Summary: Mathematics is essential in modeling and predicting a bow's performance.

Archery is the practice of propelling an arrow with a bow for the purpose of hunting, warfare, or sport. A bow is a pair of elastic limbs connected at the tips by a string. A bow acts as a spring and stores in the limbs the energy applied by the archer. As the archer releases the string, the arrow is propelled with a force proportional to the tension on the string. The path of the arrow is a parabola whose shape is determined in part by the angle of release from the bow, measured with reference to the ground.

The origins of archery are lost in the beginning of civilization and probably will never be determined with precision. The earliest bows known today were found in the Holmegaard area of Denmark and were made of elm and yew. The Holmegaard bows date form the Mesolithic period (10,000–3000 b.c.e.); however, there is archaeological evidence of projectile wounds—possibly caused by bows—from the Upper Paleolithic (40,000–10,000 b.c.e.) in all continents. It is speculated that archery was first used for hunting and, later, for warfare as social structures became increasingly complex. By the twelfth century b.c.e., archery was a decisive branch of military power. For example, the wall of the Theban temple of Ramses III depicts the Aegean fugitive fleet—driven from Crete by the Greek immigration—engaging in and losing a battle against the Egyptian fleet, whose primary weapon is shown to be archery. Archery remained the weapon of choice in the West for distance combat until the introduction of gunpowder toward the fourteenth century c.e. Archers in the medieval era would fire in a high arc, achieving accuracy by volume rather than deliberate aim. Today, archery is practiced as a precision sport and for hunting. Men's archery was one of the events of the second modern Olympics in 1900. The first Olympic archery event for women was held in 1904.

An archer's body should be perpendicular to the target and his or her feet should be shoulder-width apart. (Photos.com)

Mathematical Modeling of Bows

Since the 1930s, engineers and scientists have studied the design of bows. In 1947, C. N. Hickman made the first accurate mathematical model for flat bows, consisting of an idealized representation of two linear elastic hinges and rigid limbs with point mass (an idealized representation of a body used to simplify calculations) at the tip. More recent modeling efforts by B. W. Kooi and C. A. Bergman consider the limbs as beams that store elastic energy by bending.

The Bernoulli-Euler equation, named for Daniel Bernoulli and Leonhard Euler, describes the change in the curvature of a beam as a function of the "bending moment" (tendency to rotate about an axis) and is used to estimate the force in the string. When the archer draws the bow, the force exerted at the middle of the string causes an increase in the bending of the limbs, thus increasing the momentum and storing more energy for the shot. The elasticity modulus of the bow's material—the proportionality constant that relates limb deformation versus energy stored—determines the force with which the limbs recover their original shape after being deformed.

Mathematical modeling is a viable alternative for the evaluation of the performance of old bow models. As time passes, environmental conditions and natural processes cause considerable degradation within the cell structure of the wood used in ancient bows, which prevents a realistic assessment of the original density of the material and precludes direct testing.

Further Reading

Grayson, Charles E. *Traditional Archery From Six Continents.* Columbia: University of Missouri Press, 2007.

Kooi, B. W., and C. A. Bergman. "An Approach to the Study of Ancient Archery Using Mathematical Modelling." *Antiquity* 71, no. 271 (1971).

Miller, Frederic P., Agnes F. Vandome, and John McBrewster, eds. *History of Archery.* Beau Bassin, Mauritius: Alphascript Publishing, 2010.

Slater, Steven. "The Physics of a Wooden Bow: In Traditional Archery, Not All Bows Are Equal." *Suite 101* (July 6, 2009). http://www.suite101.com/content/the-physics-of-a-wooden-bow-a130234.

Soar, Hugh D. H. *The Crooked Stick: A History of the Longbow.* Yardley, PA: Westholme Publishing, 2009.

Juan B. Gutierrez

Ballet

Category: Arts, Music, and Entertainment.
Fields of Study: Communication; Geometry; Representations.
Summary: Ballet uses geometry to create captivating moving art.

Ballet can be considered mathematics in motion from basic counting (keeping time with music, and doing *demi-pliés* in childhood dance classes); making lines, angles, and geometric shapes in space via basic positions and choreographed routines of principal dancers and the *corps de ballet*; communicating stories in ballet productions (like the classic *Swan Lake*, or a seasonal favorite *The Nutcracker*); conversing visually among dancers (as in a *pas de deux* with Margot Fonteyn and Rudolf Nureyev); and by representing general emotions, moods, and abstract themes (as in George Balanchine's *Serenade*). Words from the French language may be common in ballet terminology, but concepts from mathematics abound as well. These representations, communications, and geometric creations can all be achieved and evidenced through the dance figures and ballet movements.

Ballet distinguishes itself from many other forms of dance through its use of the "turnout" (an outward rotation of the legs in the hip sockets to form a 180-degree line with the feet in first position). This turnout gives the dancer a strong base and the ability to move in any direction while allowing a more open body presentation to the audience, yet holding the graceful curves and shapes of the dancer's body to preserve a svelte "line." Other standard positions of the feet, carriage of the arms, or basic movements of the body produce angles such as a 135-degree arabesque, a 90-degree attitude, or a 45-degree *battement tendu*. The *rond de jambe à terre* or *en l'air* utilizes circular movements of the leg to trace semicircles or arcs, on or off the ground. These geometric lines, circles, and angles continue when basic steps become building blocks to more complicated movements. Meanwhile, dancing on the tips of the toes (*en pointe*), another distinctive ballet feature, heightens the dancer's lines in a vertical fashion. The linear extension, from head to toe, fingertip to fingertip, does not end at the extremities but continues as if through an imaginary line into the space around the dancer.

A dancer's linear extension does not end at the extremities but continues as if there were an imaginary line projected into the space around the dancer. (Photos.com)

Ballet as Geometry

One of the earliest ballet performances was the sixteenth-century *Le Balet Comique de la Reine* by Balthazar Beaujoyeulx, commissioned by the court of France. During that elaborate production, the dancers performed dozens of geometric figures involving triangles, circles, and squares for their geometric proportions and spatial configurations. These beginning ballets were influenced by the writings of Pythagoras and Plato and represented the cosmic and heavenly significance of numbers and geometry. A twentieth-century choreographer, Frederick Ashton, however, was inspired by mathematics for its sheer beauty in his creation, *Scènes de Ballet*. Working from a book of Euclid theorems, he specifically used geometry to create floor patterns and dance movements that could be viewed from any angle to see the geometric figures and "symmetrical asymmetries." Combined with the strong rhythms and counts of Igor Stravinsky's music, and geometrically patterned costumes and set details, Ashton's work was said to have beautifully combined mathematics and ballet for its visual imagery.

Notation Systems

To preserve these choreographed works of art, dance notation systems were created to symbolically represent the positions, steps, and movements of the dancers. Early seventeenth- and eighteenth-century systems, such as Feuillet notation, recorded mainly floor patterns and feet positions, whereas the twentieth-century notation systems, Labanotation and Benesh Movement Notation (written on vertical and horizontal staffs, respectively), corresponded to the scores of accompanying music. These notation systems detailed the entire body movements from head to toe of every dancer. Even with the advent of video recording, it is these symbolic notations showing graphical representations of the step details that best preserve ballets for future generations.

Further Reading

Cooper, Elizabeth. "Le Balet Comique de la Reine: An Analysis." http://depts.washington.edu/uwdance/dance344reading/bctextp1.htm.

Greskovic, Robert. *Ballet 101: A Complete Guide to Learning and Loving the Ballet.* Milwaukee, WI: Limelight Editions, 2005.

Minden, Eliza Gaynor. *The Ballet Companion.* New York: Fireside, 2005.

Schaffer, Karl, and Erik Stern. "Math Dance Bibliography." http://www.mathdance.org/MathDance-Bibliography.pdf.

Thomas, Rachel. "Scènes de Ballet." http://plus.maths.org/issue24/reviews/ballet/.

Elizabeth A. McMillan-McCartney

Ballroom Dancing

Category: Arts, Music, and Entertainment.
Fields of Study: Communication; Geometry; Representations.
Summary: Ballroom dancing allows students to approach mathematics in a variety of ways.

Ballroom dancing, considered sophisticated for its elegance, is a style of choreographed dance showcasing not only the dancers' technical skill but also their poise and style. Originally danced primarily at balls for the social elite, ballroom dancing has become a competitive sport. Dancing allows students to approach mathematics in a variety of ways, from the basic arithmetic of the beats per minute (bpm) to the geometric spatial relationship with respect to the other dancers. Choreographers Erik Stern and Karl Schaffer have created a dance called a "math dance." The purpose is twofold: to use mathematics to create dance, and to help students learn mathematics concepts through the movements of the dance. Some of the topics explored in math dances are the mathematics of rhythm, polyhedra, symmetry, and dissection puzzles.

History

The phrase "ballroom dancing" derives from the Latin word *ballare* meaning to sweep or to dance. Now considered historical dances, the original forms of ballroom dancing included the minuet and quadrille. Some steps performed in the quadrille, such as the *entrechats* (crossing the legs one in front of the other multiple times) and the *ronds de jambes* (circular movement of the leg while it is extended, toe pointing to the floor), have disappeared from the modern ballroom yet still exist in the ballet world.

In the early 1800s, the waltz made its appearance; the distance between dancing partners was considered scandalous at the time since the waltz required the partners to dance in close proximity. The early 1900s brought the birth of jazz and new dance styles as dancers moved together yet independently of each other. In addition, lively dances such as the Foxtrot, otherwise known as the one-step or two-step, moved away from the traditional placement of feet being turned out and instead called for dancers to have their feet parallel to each other. While many people are unfamiliar with any ballroom dances besides the waltz, competitive ballroom dancing has gained notoriety; it has been showcased on the ABC television show *Dancing with the Stars* and has become an Olympic sport as well.

Beats

Ballroom dancing consists of a series of dance moves, where more complicated dance steps are called "figures" or "dance figures." Each of the formally named dances has a variety of dance moves that can be put together to form a personalized performance. Determining the dance moves to use involves more than merely counting the beats. One can calculate the total number of beats that will occur in a song and then determine how many different dance moves would be necessary. For example, if one hears 12 beats in a five-second segment of the song, it can be calculated that the song has 144 bpm. If the song is exactly two minutes long, one can calculate there are 288 beats to work with for the whole song (2×144=288). Since each dance move is typically 8 beats, dividing 288 by 8 beats indicates one needs 36 dance moves. The moves can be repeated, using, for example, 9 moves 4 times each or 11 moves 3 times each (the second option gives the dancer three fewer moves than needed, requiring a dramatic flourish to end the dance). The total number of beats combined with the thematic moves of a particular dance and an individual's personal signature steps form a composite whole.

Rhythm

One rhythm option for the American-style Foxtrot consists of *Slow, Quick, Quick*, or half, quarter, quarter in 4/4 time; this approach to the dance gives teachers the opportunity to teach fractions to students using dancing. By creating a dance of successive moves in which two basic steps make one whole move, students will use fractions—adding and subtracting in 4/4 time and introducing the family of fractions

$$\frac{1}{16}, \frac{1}{8}, \frac{1}{4}, \text{ and } \frac{1}{2}.$$

This also can be done in 6/8 time with $\frac{1}{2}, \frac{1}{3}, \frac{1}{6}$, and so on.

Geometry

As the lead dancer gauges the couple's location within the coordinate plane of the dance floor, he or she keeps them spatially equidistant from other couples. In addition to the symmetry involved in the various dance moves on the dance floor, symmetry is considered within each dancer's pose and posture (the form created by the two partners together—symmetrical or asymmetrical). This symmetry can lead to an understanding of angles and curves when various dance poses are examined, and allows students the opportunity to solve problems kinesthetically when they attempt to form a mirror image of their partner while executing the dance moves.

Further Reading

Hackney, Madeleine. "Dancing Classrooms Enhance Math Skills." *Connect* 19, no. 4 (2006).
International Dance Sport Federation. http://www.idsf.net.
National Dance Council of America. http://www.ndca.org.
Watson, Anne. "Dance and Mathematics: Engaging Senses in Learning." *Australian Senior Mathematics Journal* 19, no. 1 (2005).

World Dance Council. "Welcome." http://www.wdcdance.com.

Deborah L. Gochenaur

Baseball

Category: Games, Sport, and Recreation.
Fields of Study: Data Analysis and Probability; Measurement.
Summary: Baseball is a mathematically rich sport, especially with regard to its array of statistics.

Though America's favorite sport for more than a century, the game of baseball has undergone many changes, many in response to statistics gathered regarding all parts of the game. At first, the statistics were limited to scorecard data but have expanded to include every action and detail of the game. More so, this gathering and analysis of data has expanded beyond the realm of statistical analysis, as mathematics is now used to examine all aspects of baseball—the physical characteristics and performance of its players, the analysis and modeling of each element (hitting, fielding, pitching, strategies), and the combined geometry and physics surrounding the game.

Although some fans object to this intrusion of mathematics into a competitive sport, most accept or even depend on the mathematical aspects as enriching their enjoyment of the game itself. That is, mathematics has become the arbiter in arguments, the stimulus for "hot stove league" discussions, a tool to help identify either patterns of team strengths and weaknesses or optimal strategies, and a decision-making tool for gamblers and fantasy league participants.

Sabermetrics

Bill James, a baseball writer, historian, and statistician, gave authenticity to the use of statistics in analyzing all aspects of baseball through his pioneering mathematics and statistics work. Multiple editions of his *Baseball Abstract* in the 1980s changed not only the play of the game itself but also how it is viewed by fans, and are the predecessor to many modern Web sites dedicated to analysis of the sport. James revolutionized the way mathematics is used to analyze sports to determine

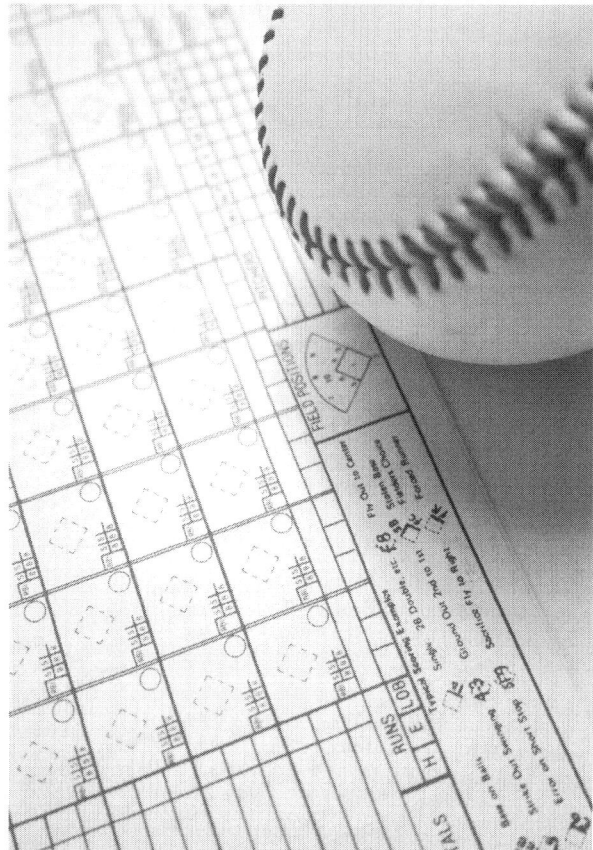

Although some fans object to using mathematics in baseball, some enjoy it as much as the game itself. (iStockphoto)

why some teams win and others lose. He coined the term "sabermetrics," which is derived from the Society for American Baseball Research acronym SABR, for his analytical and modeling methods. In 2006, *Time* magazine named him one of the most influential people in the world.

Mathematical statistics provide perspectives that explain game occurrences, provide comparative rankings of teams and players, and assist in managerial decision making. The primary example is the simple use of ratios, means, and medians as both descriptive and inferential statistics for a player, position, game, season, or career. Some examples include the following:

- Batting average, slugging percentage, on-base percentage, and batter's run average
- Effect of artificial turf on numbers of ground ball hits or base stealers' performances

- Performance of hitters and pitchers in different environments (outdoor versus dome stadiums; night games versus day games)
- Expected strike zones for umpires, given a pitcher or batter is right- or left-handed

Going beyond these descriptive statistics, the game of baseball can be analyzed using very sophisticated techniques. Some examples include the following:

- Connections between a player's characteristics and training regimens relative to game performance, or even to document the effects of steroid use
- Trend analysis, based on either a player's or team's performance (hitting, pitching, fielding) over the past five, 10, and 15 games
- Importance of pitcher throwing a "first strike"
- Effects of bringing in the infield when the bases are loaded with less than two outs
- Team winning tendencies based on run differential in innings seven, eight, and nine
- Impact of rule changes on pitching and hitting, such as the effects of elevating the pitching mound or changing foul-line distances to outfield fences
- Determination of coaching strategies such as sacrifice bunts, pitch-outs, stealing home, intentional walks, shifts of fielders for certain hitters, or use of pinch hitters and relief pitchers
- Determining the "best" all-time player in a particular position (for example, centerfielder, hitter, relief pitcher, base-stealer)
- Selection of players by professional teams during annual drafts, using both historical data for each player's performance and physical data
- Use of statistical data during contract negotiations between a player and management, or even the release or trading of players based on team needs

Mathematical probabilities, odds, and expected values can help examine the chances of particular events happening within a game or across games:

- Probability that the World Series will go four, five, six, or seven games
- Use of odds to determine personal or professional betting strategies
- Use of conditional probabilities to determine lineups or use of pinch hitters, reflecting the probability of a batter getting a hit given that the pitcher is right- or left-handed
- Correlations between a team's wins per season and player payrolls, or pitcher salaries and their ERAs
- Probability of a record being broken, either by a team or player, such as Joe DiMaggio's 56-game hitting streak

Though difficult to implement practically, geometry, trigonometry, and calculus can shed light on other important ideas:

- Length of a home run
- Actions of different pitches such as a curve ball, slider, fastball with movement, or forkball
- Determination or alteration of a hitter's batting stance or position in the batter's box
- Use of angles in fielding balls off outfield walls

Game theory also is used as part of the decision-making process within a baseball environment, leading to choices of optimal tactics. Some specific decisions are as follows:

- A manager's choice of batting lineups and pitching moves, relative to the opposing manager's choices
- A manager's calling for shifts of fielders, pitch-outs, or steals at times within a game
- A manager trying to argue, influence, or reverse decisions by umpires
- A manager's use of techniques to motivate specific players
- A team's selection of players during a draft, dependent on the player's apparent abilities, the inferred needs of other teams, and the specific draft round
- Contract negotiations involving players, agents, and team management

Finally, using all of these statistical data and mathematical modeling techniques, one can create realistic simulations of baseball games or end-of-year series, possibly using computer animations.

At the collegiate and professional levels, managers are increasingly using mathematics to remain competitive, even hiring mathematical statisticians as important parts of their staff. However, some authors and fans suggest that the team with the best players and managers will usually win, despite any use of sophisticated mathematics.

Further Reading

Albert, Jim, and Jay Bennett. *Curve Ball: Baseball, Statistics, and the Role of Chance in the Game.* New York: Springer-Verlag, 2001.

Cook, Earnshaw. *Percentage Baseball.* Cambridge, MA: MIT Press, 1966.

Eastway, Rob, and John Haigh. *Beating the Odds: The Hidden Mathematics of Sport.* London: Robson Books, 2007.

Ross, Ken. *Mathematician at the Ballpark: Odds and Probabilities for Baseball Fans.* New York: Pi Press, 2004.

Schell, Michael. *Baseball's All-Time Best Hitters: How Statistics Can Level the Playing Field.* Princeton, NJ: Princeton University Press, 1999.

Schwarz, Alan. *The Numbers Game: Baseball's Lifelong Fascination with Statistics.* New York: St. Martin's Press, 2004.

Jerry Johnson

Basketball

Category: Games, Sport, and Recreation.
Fields of Study: Data Analysis and Probability; Geometry; Measurement.
Summary: Play can be analyzed geometrically and probabilistically to inform strategy or construct simulations.

Basketball is an international sport that can be enjoyed either as a participant or as a spectator, regardless of one's sex or one's age. A growing number of coaches, reporters, and ardent fans are using mathematics to examine all aspects of basketball—the physical aspects and performance of its players, the analysis of each element (shooting, defense, strategies) of the game, and the combined geometry and physics surrounding the game. Perhaps as expected, this mathematical analysis can have opposite effects, either enriching or ruining the sports experience itself.

Basketball was intended to be a dynamic, fair competition between two teams; however, mathematical concepts and techniques can be used in a basketball environment to identify patterns of strengths and weaknesses, suggest optimal strategies for coaches and players, stimulate discussions, and resolve arguments. Statistician Dean Oliver is a well-known contributor to the statistical evaluation of basketball, which is called APBRmetrics. The name comes in part from the Association for Professional Basketball Research (APBR). This methodology is a very similar to the analysis of professional baseball using sabermetrics. Though difficult to implement practically, geometry, trigonometry, and calculus can shed light on these important ideas:

- Given a player's height, the best angle and velocity for shooting a basketball, assuming the intent is to have the basketball's parabolic arc pass through the basket (often called the "Shaq phenomena")
- The connection between the angle of shooting a ball and the event known as an "all-net" basket
- The connection between a player's height and where a player should aim a shot—at the center of basket, the front of the rim, or the back of the rim
- Use of angles in making bounce passes
- Determining defensive positions that maximize centers of gravity
- The connection between a player's position on the court and decisions to bank the basketball off the backboard as the best shot
- Comparison of the merits of shooting a free-throw underhand versus overhand

By gathering and analyzing the myriad of available data provided by a game experience, mathematical probabilities can help examine the chances of particular events happening within a game, including the following:

- The likelihood of a player making 0, 1, or 2 points in a 1-and-1 free throw opportunity
- The reality of a player having a "hot-hand," based on his or her making successive shots
- The decision as to which player should be purposely fouled at the end of a close game
- The evaluation of a player's performance in terms of "per-possession efficiency"
- The probability of a record being broken, either by a team or a player

Similarly, the collection and organization of mathematical statistics can provide perspectives that explain game occurrences, provide comparative rankings of teams and players, and assist in future decision making by coaches and team management. The usual sources of statistics are data regarding shooting, rebounding, free throws, turnovers, defensive gains, and time management. Some specific examples include the following:

- The simple use of ratios, means, and medians as descriptive statistics for a player, a position, a game, or a season
- Connections between a player's characteristics and training regimens relative to game performance
- Trend analysis, based on either a player's or a team's performance in specific ways over the past five, 10, and 15 games
- Winning tendencies based on connections to lead changes during a game or knowledge

Mathematics can be used to examine the physical aspects and performance of players, the analysis of each element of the game, and the combined geometry and physics surrounding basketball. (Photos.com)

of the team leading at the end of the third quarter
- The impact of rules changes on scoring and defenses within the sport itself, such as the observed effects of expanding either the three-point arc or the free-throw lane
- Determining the "best" all-time player in a particular position (for example, center), at a particular time in a game (for example, last-second shot), or in an era
- The seeding and selection of teams in a bracketed tournament, possibly as part of a betting pool with stated odds
- Selection of players by professional teams during the annual draft, using historical data for each player's performance in conjunction with physical data
- The use of statistical data as part of contract negotiation between players and management
- The release or trading of players based on team needs

The ideas of mathematical game theory have been applied to the decision-making process within a basketball environment, leading to choices of optimal tactics. The specific decisions range considerably:

- A coach's choice of designed offenses and defense strategies, relative to the opposing coach's choices
- A coach's calling of time-outs at opportune times within a game
- A coach trying to influence or reverse decisions by game officials
- A coach's use of techniques to motivate specific players
- A team's selection of players during a draft, dependent on the player's apparent abilities, the inferred needs of other teams, and the specific draft round
- Contract negotiations involving players, agents, and team management

Finally, using all of these available statistical data and mathematical modeling techniques, one can create realistic simulations of basketball events, full games, or even tournament series. At the collegiate and professional levels, coaches are increasingly using mathematics to remain competitive, even hiring mathematical statisticians as important parts of their staffs. Some mathematicians are even found on the court.

Retired San Antonio Spurs player Michael Robinson earned a bachelor's degree in mathematics from the U.S. Naval Academy, and is considered by many to be the best basketball player that school has ever seen. However, there are still some authors and fans who suggest the team with the best players and coaches will usually win, despite the use of sophisticated mathematics.

Further Reading

Bennett, Jay, and James Cochran. *Anthology of Statistics in Sports*. Philadelphia: Society for Industrial and Applied Mathematics, 2005

De Mestre, Neville. *The Mathematics of Projectiles in Sport*. Cambridge, England: Cambridge University Press, 1990.

Friedman, Arthur. *The World of Sports Statistics: How the Fans and Professionals Record, Compile, and Use Information*. New York: Athenaeum, 1978.

Oliver, Dean. *Basketball on Paper: Rules and Tools for Performance Analysis*. Washington, DC: Brassey's, 2004.

JERRY JOHNSON

Basketry

Category: Arts, Music, and Entertainment.
Fields of Study: Algebra; Geometry; Number and Operations.
Summary: Basket shapes and patterns are created by varying the weave.

Baskets are woven containers made of plant or artificial strips, such as palm fronds, willow branches, or fabric. People were already making baskets at least 10,000 years ago. Historians conjecture that basketry played a major role in the development of pattern, structure, and number in human cultures. Early humans could have observed birds and animals that wove to first learn the craft. Tracing basket-weaving patterns through cultures assists in creating models of human migration. "Underwater basket weaving," which is a technique in wicker, is a humorous idiom describing academic courses with low education standards or very narrow specializations.

Mathematicians study and model patterns found in baskets from around the world, including those created by the Hopi people of the southwestern United States, various African peoples, and Pacific Islanders.

Weaves

There are several types of basket weaves, each with infinitely many possible patterns. Coiled baskets are made with two types of fiber—one thick, and one soft and pliable. The thick cord or vine forms the coil. Flat, pliable strips of materials such as grass or fabric are wound a number of times around the cord, then a number of times around its previous row or several rows in the coil, connecting the rows. Craftspeople can change patterns and shapes of baskets by varying these weave numbers. Wicker is a type of basket weave consisting of relatively stiff fibers of two types. One material, the foundation, is completely rigid, and the other, the weft, is more pliable. The pattern of individual weft fibers going over and under the foundation spokes determines the look of the basket's surface. Such patterns can become very complex. Weft fibers are often soaked to make them soft during weaving.

Twining also requires rigid foundation fibers and pliable weft. Several strands of pliable fiber—usually two—go around a foundation spike on either side, cross or twist in the middle, then go around the next spike. Twining patterns are created by changing the number of wefts or formulas of skipping spikes, and introducing braiding between spikes.

Plaited baskets consist of pliable fibers woven over and under one another, typically at right angles. This weave is very similar to how woven textiles are made, and some historians believe that textiles originated from this type of basket. Formulas—whose variables are the number of fibers that go over and under in each row—determine the pattern.

The physical properties of baskets are determined by the weave, the materials, and the pattern. Wicker baskets can be very sturdy, and wicker has been used in making fences, houses, and furniture like baby cradles.

Shapes

Baskets take a variety of three-dimensional shapes, such as cylinders, cubes, and prisms. Properties of weaving often determine the shape. For example, the stiff foundation fibers of twined or wicker baskets are usually straight lines, which only allows so-called ruled surfaces. By definition in analytic geometry, ruled surfaces are generated by straight lines. Cylinders, prisms, and cones are ruled surfaces and can be made by wicker. Spheres cannot be made out of straight lines, but spherical baskets are made by coiling, plaiting, or using bendable foundations in wicker and twining. Mathematicians and mathematical artists who use basket weaving to create striking sculptured models of complex surfaces have to select appropriate weaving techniques for their projects.

Basket Patterns

Patterns found in baskets come up in many areas of mathematics. Frieze patterns have translational symmetry along lines, and there are seven types of them, all of which appear in traditional basket making. They are a part of more general wallpaper groups, of which there are seventeen types.

Some mathematics historians observed differences in patterns that involve six-fold symmetry, such as honeycombs, and more complex five-fold symmetry that comes up in basket weaving. For example, the traditional woven Malaysian ball is similar to the modern soccer ball (also known as "football") in that it contains pentagons.

Further Reading

Gerdes, Paulus. *African Basketry: A Gallery of Twill-Plaited Designs and Patterns.* NP: Lulu Publishing, 2008.

University of East Anglia. "Basket Weaving May Have Taught Humans to Count." *ScienceDaily* (June 8, 2009). http://www.sciencedaily.com/releases/2009/06/090604222534.htm.

Zaslavsky, Claudia. *Multicultural Mathematics: Interdisciplinary Cooperative-Learning Activities.* Portland, ME: Walch Education, 1993.

Maria Droujkova

Betting and Fairness

Category: Games, Sport, and Recreation.
Fields of Study: Algebra; Data Analysis and Probability.
Summary: Mathematics is used to analyze betting and probabilities for games of chance and for investing in the stock market.

A pivotal moment in the early development of probability occurred in 1654, as the French mathematicians Blaise Pascal and Pierre de Fermat exchanged a series of letters. Pascal and Fermat were wrestling with questions involving the fair payoff for a gambler who is forced to quit in the middle of a game. In modern language, they were calculating the "expected value" of the game's payoff (the average payoff under the various possible outcomes, weighted according to the likelihood of those outcomes). A bet is said to be "fair" if the price of placing it is equal to the expected value of the payoff. Betting plays an integral part in our modern society. People place bets in casinos and at sporting events, as well as by buying lottery tickets. They are also placing bets when purchasing insurance or investing in the stock market. Some of these bets are fair, some are unfair, and some cannot be objectively categorized.

The primary problem that Pascal and Fermat solved (each employing a different method) can be used to illustrate some important ideas on fairness. In the problem, two gamblers are playing a game in which a coin is repeatedly tossed. The game is interrupted at a point where 2 more heads are required for Player A to win and 3 more tails are required for Player B to win (whichever occurs first). How should the potential winnings be divided at this stage of the game?

Fermat solved the problem by observing that at most 4 tosses remain in order to identify the winner, and that there are 16 equally likely ways in which 4 tosses could occur:

HHHH, HHHT, HHTH, HTHH, THHH, HHTT, HTHT, HTTH, THHT, THTH, TTHH, HTTT, THTT, TTHT, TTTH, and TTTT.

In 11 of these possibilities (the first 11 items on the list), Player A would win, because 2 heads occur before 3 tails; in the other 5 possibilities, Player B would win, because 3 tails occur first. Therefore, Fermat reasoned that Player A should receive 11/16 of the winnings, and Player B should receive 5/16 of the winnings. In modern language, Player A would win the game with probability 11/16 and Player B would win with probability 5/16; Fermat was calculating the "expected value" of the winnings for each player.

Suppose that up to this point in their game, neither Player A nor B has paid any money for the opportunity to play, but that they are now required to pay a total of $1, altogether, and that this dollar will constitute the winnings. Fermat's solution to the previous problem allows for a fair method of dividing the payment: Player A should pay 11/16 of the dollar and Player B should pay 5/16, so that the payments match the expected winnings. In other words, if the game is being played for a $1 payoff, then the price for a fair bet is 11/16 of a dollar for Player A and 5/16 of a dollar for Player B.

Lotteries and Casinos

State-run lotteries are unfair to the player who purchases a ticket, because some of the revenue goes to the state and is not redistributed to the winner(s). Of course, even if all of the ticket revenue were paid to the winner(s)—so that the bets were fair—a lottery would be unfavorable to almost every player. Nonetheless, lotteries attract large numbers of players because people are willing to pay a small amount for the minuscule chance of winning a fortune.

A similar motivation attracts bettors to casinos, where almost all games are unfair. This casino advantage is known as the "house edge." On average, the house edge at a casino is 2% to 3%, which means that for each dollar that is bet, the house makes a profit of 2 or 3 cents. Over thousands of bets, this adds up to a significant profit. Some games, like slot machines, can have a house edge of up to 15%. Typically (in roulette, slot machines, and craps, for instance), the odds for each bet are slightly in favor of the house. Blackjack is a rare example of a casino game in which a player might be able to place bets that are better than fair from the player's perspective. In blackjack, two initial cards are dealt to each player as well as to the dealer. Certain strict rules dictate whether additional cards are dealt to the dealer, while each player has the choice of whether to receive additional cards. The objective of each player is to hold a total card value closer to 21 than the dealer holds, without going over. Each player knows which cards he or she holds, as well as some of the cards held by the dealer and other players, since some cards are dealt face up. An adept player can also keep track of cards that have been used in previous games following the last shuffle—though casinos often dissuade such card counting by combining several decks and shuffling regularly. By using this information, it is possible for a player to calculate the probability of drawing a particular card and, therefore, the expected value of the payoffs under the options of either receiving an additional card or not; often, one of these expected values is greater than the amount of the bet.

Subjective Probabilities

Early in the twentieth century, mathematicians realized the need to define probability in a rigorous way, if it were to be a formal part of mathematics. In problems involving tossing fair dice or coins, or counting card hands, it was obvious what should constitute the probabilities of the various occurrences, but in many other situations it was unclear. Usually people thought of probabilities as idealized frequencies: if a fair coin is tossed many times, for example, then the fraction of tosses which land heads should be approximately 1/2; so a fair bet for a $1 payoff on heads should cost $0.50. But there is not an obvious analogy for two boxers, for example, about to fight a match. Also, probability was becoming an increasingly important tool for the physical sciences, and mathematical theorems were required. As such, an axiomatic system was necessary. The Russian mathematician A. N. Kolmogorov and the Italian philosopher and mathematician Bruno de Finetti independently provided such a framework in the 1930s. Although different in appearance, their definitions are equivalent in most situations.

De Finetti's concept of a probability stems from gambling: the probability of an event is the price for a $1 payoff bet on that event. These prices may be assigned in whatever way one wants (hence the label "subjective probabilities"), provided certain consistency conditions are met. For example, suppose even money is coming into a betting house on two teams preparing to play a baseball game. This indicates that the bettors collectively value the two teams as equally likely to win the game. Ignoring the house fees, the price for a $1 payoff bet on either team is $0.50, because after the game, the entire pool of money will be redistributed to those who bet on the winning team.

Suppose, however, a particular bettor favors the home team, believing that team to have a 3/4 probability of winning the game. Then this bettor would price a $1 payoff bet on that team at $0.75; for this bettor, the $0.50 price generated by the betting pool is a bargain. From this bettor's perspective, a bet on the home team is better than fair: the price for a $1 payoff bet is $0.50, but the expected value of the payoff is $0.75. Such situations occur beyond sporting events, perhaps most prominently in the stock market. The fact that individuals' valuations often differ from those of the collective public is the driving force behind the trading of stocks. Individuals buy stocks that they believe to be undervalued and sell stocks that they believe to be overvalued. Because they are predicting the future performance of these stocks, they are essentially placing bets that they believe to be better than fair. In 1956, John Larry Kelly, Jr., a physicist who worked at Bell Labs, formulated and described the Kelly criterion. This algorithm for determining an optimal series of investments (or bets) is based on probability and economic utility theory, which tries to mathematically quantify satisfaction. In recent years, the Kelly criterion has been incorporated into many mainstream investment theories and betting strategies.

Further Reading

David, F. N. *Games, Gods and Gambling: A History of Probability and Statistical Ideas*. New York: Dover Publications, 1998.

Devlin, Keith. *The Unfinished Game: Pascal, Fermat, and the Seventeenth-Century Letter That Made the World Modern*. New York: Basic Books, 2008.

Epstein, Richard A. *The Theory of Gambling and Statistical Logic*. 2nd ed. San Diego, CA: Academic Press, 2009.

Packel, Edward. *The Mathematics of Games and Gambling*. 2nd Ed. Washington, DC: The Mathematical Association of America, 2006.

Von Plato, Jan. *Creating Modern Probability*. New York: Cambridge University Press, 1994.

John Beam

Billiards

Category: Games, Sport, and Recreation.
Fields of Study: Geometry; Number and Operations.
Summary: Playing billiards depends on an understanding of spin, momentum, and angles.

Billiards is a cue sport game that involves the use of a rectangular table, billiard balls, and a stick called a "cue." Mathematics and physics are two important components of playing the game well. There are many different games within the cue sports that Americans typically name "billiards." Billiard tables with pockets comprise games that are termed as "pool" or pocket billiards. The rectangular table has two long sides (twice the short side) and two short sides with six pockets—one at each corner, and one midway along the longer two sides of the table. The object of the game is to hit the billiard balls into the pockets using a cue ball (the lone white ball in the set). Gaspard Coriolis, known today for the Coriolis effect, wrote a work on the mathematics and physics of billiards in 1835. He stated that the curved path followed by the cue ball after striking another ball is always parabolic because of top or bottom spin. Further, the maximum side spin on a cue ball is achieved by striking it half a radius off-center with the tip of the cue.

The game of billiards is also a source for interesting mathematical problems, which are connected to dynamical systems, ergodic theory, geometry, physics, and optics. In mathematical billiards, the angle of incidence is the same as the angle of reflection for a point mass on a frictionless domain with a boundary. The dynamics depend on the starting position, angle, and geometry of the boundary and the table. Mathematicians investigate the motion and the path of the ball on a variety of differently shaped flat and curved tables, like triangular or elliptical boundaries or hyperbolic tables. In 1890, mathematician Charles Dodgson, better known as *Alice in Wonderland* author Lewis Carroll, published rules for circular billiards and may have also had a table built. In 2007, mathematician Alex Eskin won the Research Prize from the Clay Mathematics Institute for his work on rational billiards and geometric group theory.

Eight Ball

Eight ball is the pool game most commonly played in the United States, and it involves 16 billiard balls. To

Billiard players can use transformational geometry to try to hit the ball so that it will return to a pocket. (Photos.com)

one-quarter of the long side known as the head spot. The point of the triangular-shaped racked set of billiard balls is placed on the opposite short end at one-quarter of the length of the long side from the other short side and is known as the foot spot. After one player "breaks" by hitting the cue ball from the head spot into the racked set of balls, the player then hits a set of balls into the pockets. A shot that does not cause a ball of his or her set to go into the pocket results in the next shot going to the other player.

Billiards Geometry and Physics

Shooting the balls into the pockets requires an understanding of angles and momentum, as well as placement of the cue so that the correct spin is achieved to place the cue ball where it can achieve the target ball going into a pocket. Coriolis investigated 90-degree and 30-degree rules of various shots and measured the largest deflection angle the cue ball can experience. Both skill and geometric understanding contribute to successful shots. Some shots require straight shooting; some shots need to be "banked" in by using the table sides. Players can use transformational geometry to approximate where on the table to hit the ball for it to return to a pocket. By measuring the angle from the ball to the side being used to bank off and reflecting the same angle with the cue stick, one can see the most viable spot to aim for so that the path of the caromed ball ends in a pocket. Using the diamonds found on the sides of most tables is one way of measuring these angles, and some systems for pool and billiards play use the diamonds. Using the diamond system for a different billiard game, Three Cushion Billiards is demonstrated on the 1959 Donald Duck Disney cartoon *Donald in Mathmagic Land*. The demonstration shows that it is possible to use subtraction to know where to aim the ball in relation to a diamond to make sure that all three balls are hit.

Further Reading

Alciatore, David. "The Amazing World of Billiards Physics." http://billiards.colostate.edu/physics/Alciatore_SCIAM_article_posted_version.pdf.

Tabachnikov, Serge. *Geometry and Billiards*. Providence, RI: American Mathematical Society, 2005.

Linda Hutchison

Birthday Problem

Category: Friendship, Romance, and Religion.
Fields of Study: Algebra; Data Analysis and Probability; Measurement; Number and Operations.
Summary: The Birthday Problem is a classic example of how probability can reveal counterintuitive truths.

The Birthday Problem is a classic probability problem first presented by mathematician and scientist Richard von Mises in 1939, though the fundamental combinatorial concepts involved can be traced back as far as India in the sixth century B.C.E. Today, it is one of the most-explored problems in classrooms, often introduced as early as the middle grades. The problem asks: Given that there is some number of people (n) in a room, what is the probability that at least two of them share the same birthday? One of the aspects that makes this problem so intriguing is that the answer is much different than people intuitively expect.

Solving the Birthday Problem

The extreme cases of the problem are easy to determine logically. If there are fewer than two people, then it is impossible to have two who share the same birthday, making the probability 0. If there are more people than days in a year, then at least two people must share the same birthday, making the probability 1. Von Mises assumed a fixed 365 days per year, ignoring February 29 as a possible birthday, so the probability is always 1 if there are 366 or more people. More interesting and challenging are the cases for which there are anywhere from 2 to 364 people. For the purposes of modeling and computation, it is assumed that it is equally likely that someone will be born on one day of the year versus another, so there is a

$$\frac{1}{365} \text{ chance}$$

that a person will be born on any particular day.

The Birthday Problem is solved using the mathematical ideas of permutations and combinations, and it is more easily approached if one asks a slightly different but complementary question: What is the probability that everyone in the room has a unique birthday? That is, that no one shares. If there are two people in the room, the first can be born on any of 365 days of the year and the second must be born on any of the remaining 364 days. If there are three people in the room where the first is born on a particular day, the second must be born on one of the remaining 364 days of the year and the third on one of the remaining 363 days. The probability is

$$\frac{365}{365} \times \frac{364}{365} \times \frac{363}{365}.$$

If there are four people, the probability is

$$1 \times \frac{364}{365} \times \frac{363}{365} \times \frac{362}{365}.$$

This pattern can be generalized as

$$P_{(\text{no match})}(n) = \frac{_{365}P_n}{365^n}.$$

The probability that at least two people in a room of n people share a birthday is 1 minus the probability that there is no match, which can be used to generate the following probabilities.

Number of people in the room	Probability that two of these people share a birthday
2	.00274
10	.117
20	.411
23	.507
30	.706
50	.970
60	.995

When there are 23 people in the room, there is slightly more than a 50% chance that two people will share the same birthday, which answers the original question. The probability of at least one match increases quickly and nonlinearly with the number of people, so that when the number reaches 60 (well below the certainty value of 366 people), there is a 99.5% chance that there will be a match—*almost* certain. For example, as of 2010, there are six pairs of men who share a birthday among the 74 unique winners of the Academy Award for Best Actor. For women, there are three pairs among the 69 unique Best Actress winners.

Applications of the Birthday Problem

Applications of the Birthday Problem exist in many fields. One is called Class Phenotype Probability. Given six characteristics (blood type, RH positive/negative, sex, mid-digital hair positive/negative, earlobes attached/unattached, and PTC taste receptor), it is possible to determine the probability that a particular combination exists and also the probability that two people share the same combination. This possibility is quite valuable in medicine when considering the likelihood of finding matches between donors and recipients. In computer security, a birthday attack is a computationally intensive strategy used to break encrypted digital signatures. A "collision" occurs when different sets of data yield the same cryptographic hash value, which is a function of the input data. The attack repeatedly evaluates a hash-generating function using random inputs until the output creates a collision with the true hash value it seeks to duplicate. On average, $1.2 \times \sqrt{k}$ trials are needed to get a match, where k is the number of possible outputs (for example, a 64-bit hash value has about 1.8×10^{19} outputs). The birthday attack strategy becomes much less efficient as the hash length increases.

There are interesting extensions of the Birthday Problem based on slightly altering the question or assumptions. The first comes from considering the chance that *three* or more people share a birthday (or four, or five, and so forth). The Almost-Birthday Problem expands the problem to finding at least two people whose birthdays are within one day of each other. The Movie Line Problem states that the first person in a line for a movie whose birthday matches someone in front of them wins free tickets, and it seeks to find where someone should stand to have the best chance of winning. The Goldberg Extension computes the expected number of different birthdays in a group, while the Tuesday Birthday Problem is given as, "I have two children, one of whom is a boy born on a Tuesday. What is the probability that my other child is a boy?" Other variations assume unequal distributions of birthdays throughout the year. As with the original problem, solutions usually run contrary to most people's intuition. The ideas provide the basis for many applied investigations, such as the photon behavior modeling done by mathematical physicist Satyendra Nath Bose, after whom the subatomic particle "boson" is named.

Further Reading

Borja, Mario Cortina, and John Haigh. "The Birthday Problem." *Significance* 4, no. 3 (2007).

Goldberg, Samuel. "A Direct Attack on a Birthday Problem." *Mathematics Magazine* 49, no. 5 (1976).

Mosteller, Frederick. *Fifty Challenging Problems in Probability With Solutions*. New York: Dover Publications, 1987.

Lidia Gonzalez

Board Games

Category: Games, Sport, and Recreation.
Fields of Study: Algebra; Data Analysis and Probability; Geometry.
Summary: While some games are explicitly mathematical, others are implicitly governed by math.

Humans have been playing games for as long as they have been around. Johan Huizinga was the first to call the attention to the fact that play precedes culture. Board games, a very organized form of play, are part of human social nature. Human communities may differ in many ways, but they all play games. From the ancient Mancala, practiced for millennia in Africa, to our Monopoly, we find board games in many societies. Besides their cultural relevance—they are studied by anthropologists, historians, and others—board games are characterized by their sets of rules, which show mathematical structures and connections that are at times very surprising.

Game Classifications

Chess and Go come to mind as examples of traditional board games, and Monopoly and Scrabble are examples of proprietary games. The distinction between the two types of games is not always easy to identify. In chess, the movements of the pieces and the other rules are the main considerations. Chess is an abstract game, not considering the fact that it originally emulated a battle between two armies. Chess does have similarities with other games. When playing representational games like Monopoly or Diplomacy, players find themselves focusing on the possibilities and strategic choices, forgetting the particular settings. Accoring to David Parlett, positional games refer to games where pieces are played in a board or any other set of markings, as chess, checkers, and Go, and "theme" games are generally representational and commercial, like Monopoly and Diplomacy.

Board game classification has been inspired in the fact, first noted by H. J. R. Murray, that games are typical of early activities of man—the battle, the siege, the race, the hunt, alignment, arrangement, and counting. Parlett's classification, which evolved from Murray's and others, is as follows. In race games, the board is a linear track where each player tries to be the first to reach a particular cell or remove a set of pieces from the board. Most of the games under this category use dice or other randomizing devices, like Chutes & Ladders, Ludo, and Backgammon, but not all, such as Hare & Tortoise.

Space games, typically two-dimensional and free placing, comprise the alignment games, as Nine Men Morris; connection games, as Hex and Twixt; traversal games, in which a player tries to have one or several pieces cross the board, as Breakthrough, Halma, and Chinese Checkers; configuration games, where players try to achieve certain displays with their pieces, as Agon; restriction games, where the aim is to try to block the adversary, like Pentominoes; and occupation games, in which the winner is the player who achieves more space in the board, as in Go and Othello. Chase games are asymmetrical, one player having several pieces while the other has only one or two. Their goals are also distinct, as in Fox & Geese. "Displace games" include chess and checkers, where a player aims at capturing most of his opponent's pieces (as in checkers) or a particular one (as in chess), and other war games; the family of Mancala games belongs also to this class.

History

The Royal Game of Ur, also known as the Game of Twenty Squares, was found in the south of Iraq in the 1920s and is about 4500 years old. The board shows twenty squares, 12 in a three-by-four rectangular array, six in two rows of three, and two connecting cells. The reverse of the board corresponding to the 12 cells showed a zodiac, illustrating that in the past, the same object could be a board game and a divinatory device. Two cuneiform clay tablets give the exact rules for this game. Each player had seven pieces, which moved across the board according to the toss of three tetrahedral dice.

A similar game is found in Ancient Egypt, Senet or the Game of Thirty Squares. It was a race game as well, but it was more than a simple toy. In funerary monuments that date from 4000 years ago, images are shown of the deceased playing Senet against an invisible adversary. Osiris, which is present but not shown, decides on matters of life after death.

The Royal Game of Ur and Senet can be viewed as the oldest relatives of the modern Backgammon, a game in which the moves are decided by the players upon tossing two cubic dice. The player who better understands the probability laws that rule the dice is most often the winner.

The Chinese game Go is four millennia old. Nowadays, it remains one of the most complex games, despite the simplicity of its rules. Go is played on the intersections of a 19-by-19 grid, and each player fights to control the largest area.

Pure strategy games could also be found in Ancient Greece, like Petteia. This game, and the Roman Ludus Latrunculorum, shared the shape of the board, checkered, and the orthogonal movement of the pieces.

Chess, which originated in India about 1400 years ago, traveled to the West with the Arabs, and saw its rules evolve in the process. It was originally created as a war game between two armies, and its pieces represented the actors of the battle. However, the abstract shapes that reached Europe gave way to the symbolic representation of the European medieval society.

The Arabs introduced several other games in Europe. One game they introduced, Alquerque, was played on the intersections of a five-by-five lined board. The adaptation of this game to the chessboard originated the game of Checkers.

Board Games and Mathematics

The oldest known pedagogical game is Rithmomachia, also known as Philosopher's Game. It was invented in the eleventh century as a didactical device to teach mathematics. It was practiced wherever Boethius's arithmetic was taught. Pythagorean in nature, this tradition of mathematics dominated teaching at churches and universities for more than 500 years. In an eight-by-16 board, two armies fought each other. Pieces carried numbers and could have one of three shapes: circular, triangular, or square.

The movements depended on the shape of the piece played; the captures depended on the numbers and on arithmetical calculations. Victory was attained by means of a configuration of pieces holding numbers in progression (arithmetic, geometric, harmonic, or combinations of the three). This game spread throughout Europe, and only when the mathematical curriculum at universities changed in the sixteenth century did it vanish. Losing its pedagogical goal turned out to be fatal, as Rithmomachia lacked the qualities to survive as a purely recreational activity. Chinese scholars of the eleventh century also published work on permutations based on the Go board. John H. Conway's twentieth-century research on the game contributed to the invention of surreal numbers and the development of combinatorial game theory.

Ludus Astronomorum was a board game for seven players based on Ptolemaic astrological principles. In the sixteenth century, William Fulke, a professor at Cambridge who had written a manual of the Philosopher's Game, created two other games. One, intended to improve on the astronomy game, was Ouranomachia, the other, created to teach geometry, was Metromachia. Fulke published one book on each.

In the eighteenth century, George Berkeley invented a game to help teach algebra, a subject Berkeley had in very high consideration. The game was Ludus Algebraicus and essentially functioned as a randomizing device to generate algebraic equations.

Charles Dodgson invented a game in the nineteenth century to practice logical deduction and wrote a book about it, *The Game of Logic*, under his pen name, Lewis Carroll.

In Ireland, mathematician William Hamilton created in 1857 the Icosian Game and soon after Traveller's Dodecahedron. This comprised a dodecahedron and a piece of thread that should touch every vertex according to some rules. It was this game that gave rise to the concept of Hamiltonean Graph.

The familiar game of Nim in which a move consists of choosing from one of a pile of beans and reducing its cardinality, was first solved mathematically at the beginning of the previous century. In its normal form, where the winner is the one who takes the last bean, is the paradigm of a class of games studied in Combinatorial Game Theory. The familiar children's game Dots & Boxes was also treated mathematically with the same techniques. Some traditional games, like Konane, can be approached the same way.

The game Hex was invented independently by both Piet Hein and John Nash in the 1940s. It is a connection game played on a diamond-shaped board of hexagonal cells. David Gale noted that a game of Hex can never end in a tie, and that this fact is logically equivalent to a deep theorem in topology.

Abstract games with complete information and no chance devices are also called mathematical games. The mental processes present in their practice and in a typical mathematical activity, like problem solving, are far from disjointed.

Further Reading

Avedon, Elliot M., and Brian Sutton-Smith. *Study of Games*. New York: John Wiley & Sons, 1971.

Berlekamp, Elwyn R., John H. Conway, and Richard K. Guy. *Winning Ways for Your Mathematical Plays*. Natick, MA: Ak Peters, 2001.

Huizinga, Johan. *Homo Ludens*. New York: Routledge, 2008.

Murray, Harold James Ruthven. *A History of Board-Games Other Than Chess*. New York: Hacker Art Books, 1952.

Parlett, David. *Oxford History of Board Games*. New York: Oxford University Press, 1999.

Jorge Nuno Silva

Calculus in Society

Category: School and Society.
Fields of Study: Calculus; Connections.
Summary: Since its introduction in the seventeenth century, calculus has been applied to more and more practical endeavors, from engineering and manufacturing to finance.

Since its initial development in the seventeenth century, calculus has emerged as a principal tool for solving problems in the physical sciences, engineering, and technologies. Applications of calculus have expanded to architecture, aeronautics, life sciences, statistics, economics, commerce, and medicine. Contemporary society is impacted continually by the applications of calculus. Many bridges, high-rise buildings, airlines, ships, televisions, cellular phones, cars, computers, and numerous other amenities of life were designed using calculus.

Since the 1970s, calculus in conjunction with computer technology has resulted in the emergence of new areas of study such as dynamical systems and chaos theory. Such vast applications have established the study of calculus as essential in preparation for numerous careers. Indeed, calculus is considered among the greatest achievements of humankind, making it worthy of study in its own right in a society that places rational thought and innovation in highest esteem. Recent curricular and pedagogical reforms in calculus have made it more academically accessible to the school population.

What Is Calculus?

Calculus originated from studying the physical motions of the universe, such as the movement of planets in the solar system and physical forces on Earth. It involves both algebra and geometry, in combination with the concepts of infinity and limits. In contrast to algebra and geometry, which focus on properties of static structures, calculus centers on objects in motion. There are two principal forms of calculus, differential calculus and integral calculus, which are inversely related. At its most basic level, differential calculus is used in determining instantaneous rates of change of a dependent variable with respect to one or more independent variables; integral calculus is used for computing areas and volumes of nonstandard shapes.

Who Invented Calculus?

In the late seventeenth century, Isaac Newton (1646–1727) of England and Gottfried Wilhelm Leibniz (1646–1716) of Germany independently invented calculus. Isaac Newton began his development of calculus in 1666 but did not arrange for its publication. He presented his derivations of calculus in his book, *The Method of Fluxions*, written in 1671. This book remained unpublished until 1736, nine years after his death. Gottfried Leibniz began his work in calculus in 1674. His first paper on the subject was published in 1684, 50 years earlier than Newton's publication. Because of these circumstances and fueled by the eighteenth-century nationalism of England and Germany, a bitter controversy erupted over who first invented calculus. Was it Isaac Newton or Gottfried Leibniz?

Investigators found that Leibniz had made a brief visit to London in 1676. Supporters of Newton argued that during that trip, Leibniz may have gained access to some of Newton's unpublished work on the subject from mutual acquaintances within the mathematics community. However, these two prominent and outstanding mathematicians used their own unique derivations and symbolic notations for calculus, with Newton developing differential calculus first and Leibniz developing integral calculus first. For many decades, the calculus feud divided British mathematicians and continental mathematicians, and it remains a historical mystery into the twenty-first century. It was an unusual controversy in that it erupted rather late in the development of calculus and was ignited by the respective followers of Newton and Leibniz. In the twenty-first century, the general consensus is that both Newton and Leibniz invented calculus, simultaneously and independently.

Isaac Newton (1646–1727): The Man
Isaac Newton was revered in England during his lifetime and is recognized as one of the foremost mathematicians and physicists of all time. In addition to his invention of calculus, Newton is famous for designing and building the first reflecting telescope, formulating the laws of motion, and discovering the white light spectrum. He held many prestigious positions, including Fellow of Trinity College, Lucasian Professor of Mathematics, Member of Parliament for the University of Cambridge, Master of the Royal Mint of England, and many others. Even though Newton was extremely productive and admired universally for his work, on a personal level he was humble, cautious of others, and angered by criticism. His modest nature is embodied in his famous statement, "If I have seen farther than others, it is because I stood on the shoulders of giants." His works in mathematics and physics were recognized throughout Europe when he was honored as Fellow of the Royal Society of London in 1672. He subsequently served as the Society's president from 1705 until his death. In 1705, Newton was knighted in Cambridge by Queen Anne of England for his contributions to the Royal Mint. In 1727, Newton's name was immortalized in English history by his burial in London's Westminster Abbey and by the accompanying monument honoring his contributions to mathematics and science.

Gottfried Leibniz (1646–1716): The Man
Gottfried Leibniz is recognized as one of Germany's greatest scholars of philosophy, history, and mathematics. He was the son of a philosophy professor and a leader in the philosophy of metaphysics. His optimism is reflected in his words, "We live in the best of all possible worlds." On a personal level, Leibniz was considered likeable, friendly, and somewhat boisterous. Professionally, Leibniz was employed by a succession of German princes in the capacities of diplomat and librarian. He planned and founded several academies throughout Europe. For his knowledge of law, he was appointed Councilor of Justice for the Germanic regions of Brandenburg and Hanover. Similarly, Russian Tsar Peter the Great appointed Leibniz as Court Councilor of Justice for the Habsburgs. For his work in mathematics (derivations in calculus and invention of the binary number system), in 1673, Leibniz was appointed Fellow of the Royal Society of London, a society honoring outstanding mathematicians and scientists throughout Europe. By 1706, however, Leibniz's stellar reputation had begun to disintegrate. Accusations of plagiarism regarding the invention of calculus were unrelenting until Leibniz's death in 1716. In contrast to Newton, the only mourner at Leibniz's funeral was his secretary. Eventually, more than a century after his death, Leibniz's outstanding contributions to mathematics were recognized in Germany when a statue was erected in his honor at Leipzig, one of Germany's major centers of learning and culture.

Interestingly, it is Leibniz's symbolic notations for calculus, namely dy/dx and $\int y\, dx$, that have stood the test of time. These notations are most prevalent in calculus classrooms in the twenty-first century because of their consistency with the operations of differential equations and dimensional analysis. The most significant contribution to mathematics by Newton and Leibniz was their derivations of the Fundamental Theorem of Calculus, a theorem that unites both differential and integral calculus.

Building on Newton's and Leibniz's Work
Following the invention of calculus, additional contributions to calculus were made by John Wallis (1616–1703), Michel Rolle (1652–1719), Jacob Bernoulli (1654–1705), Guillaume de l'Hôpital (1661–1704), Brook Taylor (1685–1731), Colin Maclaurin (1698–1746), Joseph-Louis Lagrange (1736–1813), Bernard Bolzano (1781–1848), Augustin-Louis Cauchy (1789–1857), Karl Weinerstrasse (1815–1897), and Bernhard Riemann (1826–1866).

The Power of Calculus
The power of calculus in contemporary society rests primarily in its applications in the physical sciences, engineering, optimization theory, economics, geometrical measurement, probability, and mathematical modeling.

The following is a sampling of basic applications using the two major branches of calculus.

Applications of Differential Calculus

- *Environmental science*: An oil tanker runs aground and begins to leak oil into the ocean and surrounding land areas, resulting in potentially devastating consequences. Differential calculus can be used to supply information essential for assessing the leakage and resolving the problem. For example, the rate and volume at which the oil is leaking can by determined using calculus.
- *Business and economics*: Important applications of calculus in business and economics involve marginal analysis (known as the first derivative). Marginal costs, revenues, and profits represent rates of change that result from a unit increase in product production. This information is valuable in developing production levels and pricing strategies for maximizing profits.
- *Medicine*: Calculus can be used for evaluating the effectiveness of medications and dosage levels. For example, calculus can be used in determining the time required for a specific drug in a patient's bloodstream to reach its maximum concentration and effectiveness.
- *Biology and chemistry*: Assessments of chemical treatments for reducing concentration levels of biological contaminants (such as insects or bacteria) can be determined by calculus. For instance, calculus can be used in measuring the concentration levels, effectiveness, and time necessary for a chemical treatment supplied to a body of water to reduce its bacterial count to desired minimal levels.
- *Physics* (*velocity and acceleration*): For moving objects (such as rolling balls or hot-air balloons), their maximum velocities, accelerations, and elevations can be determined using calculus.
- *Politics*: The number of years required in a city for the rate of increase in its voting population to reach its maximum can be determined using calculus.
- *Manufacturing*: The design of containers, meeting specific constraints, can be determined using calculus. For example, calculus will supply the dimensions of a container that will maximize its volume or minimize its surface area.

Applications of Integral Calculus

- *Inverse of differential calculus*: In mathematics, most operations have inverse operations. In calculus, the inverse of differentiation is integration. Therefore, a fundamental application of integral calculus is to find functions that produce the answers to a problem in differential calculus.
- *Measurement, area, and volume*: Integral calculus can be used to find (1) the areas between the graphs of functions over specified intervals, (2) the surface areas of three-dimensional objects, and (3) the volumes of three-dimensional objects.
- *Centroids*: The centroid (or center of mass) of an object can be found using integral calculus. For two- and three-dimensional objects, the centroid is the balancing point of the object. Calculus can be used to locate the position of the centroid on the object.
- *Fluid pressure*: Integral calculus is essential in the design of ships, dams, submarines, and other submerged objects. It is used in determining the fluid pressure on the submerged object at various depths from the water's surface. This information is essential in the design of submerged objects so they will not collapse.
- *Physics* (*work*): When a constant force is applied to an object that moves in the direction of the force, the work produced by the force is found by multiplying the force by the distance moved by the object. However, when the applied force is not constant or is variable, calculus can be used in determining the work produced by the variable force (for example, the variable force needed to pull a metal spring, or the force exerted by expanding gases on the piston in an engine).

The aforementioned applications are examples of the most elementary applications of calculus. In the technological world of the twenty-first century, applications of calculus continue to evolve. The consequences of calculus are ubiquitous in contemporary society and impact every walk of life.

Recommendations for Mathematics Curriculum Reform

In 1983, following a harsh report from the National Commission on Excellence in Education, U.S. society began to question seriously the effectiveness of its educational systems. The report, titled *A Nation at Risk: the Imperative for Educational Reform*, was commissioned by U.S. President Ronald Reagan. The report cited U.S. students for their poor academic performance in every subject area at every grade level and their underachievement on national and international scales. The Commission warned the United States that its education system was "being eroded by a rising tide of mediocrity." In the years that followed, the Commission's explicit call for educational reform in U.S. schools served to generate numerous curricular reform efforts at the pre-college and college levels.

In response to this call for reform, in 1987, the Mathematical Association of America (MAA) and the National Research Council (NRC) co-sponsored a conference held in Washington, D.C., titled Calculus for a New Century. The conference was attended by more than 600 college and pre-college calculus teachers. The conference focused on the nature and need for calculus reform in college and pre-college institutions throughout the nation. During that conference, the phrase "Calculus should be a pump, not a filter in the pipeline of American education" became a national mantra for calculus reform.

National educational assessments conducted in 1989 further supported initiatives for calculus reform. During the 1980s, approximately 300,000 U.S. college students were enrolled annually in science-based calculus courses. Of that number, only 140,000 students earned grades of D or higher. Thus, more than 50% of U.S. college students were failing the calculus courses required for their majors, which included mathematics, all of the natural and physical sciences, and computer science. These bleak statistics served to motivate concerned calculus teachers to examine the traditional calculus curriculum, as well as their own teaching methodologies, with the intention of increasing course enrollments, student achievement, and enthusiasm for the subject.

Their efforts resulted in major calculus reform initiatives as early as 1989. The first set of recommendations for reform in school mathematics (grades prekindergarten–12) came from the National Council of Teachers of Mathematics (NCTM). These recommendations were delineated in NCTM's publication, *Principles and Standards for School Mathematics* (also known as NCTM Standards).

Four overarching standards (called Process Standards) were identified for improving mathematics instruction at all levels. These standards identified problem solving, reasoning and proof, connections, and communications as the four primary foci for mathematics instruction. During the 1990s, most U.S. states adopted this document as their curriculum framework for school mathematics. Decisions regarding the mathematics curriculum, textbook selections, and instructional strategies were revised in accordance with the recommendations of the NCTM Standards. Interestingly, the same document served to inspire pedagogical reform in mathematics at the college level, especially in calculus.

Traditional Calculus Versus Reformed Calculus

Until 1990, the calculus curriculum had remained basically the same for decades. The traditional calculus curriculum reflected formal mathematical language, mathematical rigor, and symbolic precision. Computations with limits, mathematical proofs, and elaborate mathematical computations were common practice in calculus classrooms. Students took careful notes, asked clarifying questions, and completed voluminous amounts of homework in preparation for test questions similar to those completed for homework. Instruction was teacher-centered and delivered through a lecture approach. Relevant applications were seldom considered, and graphing calculators and computers were rarely used in calculus instruction, and students were not allowed to use them for computations, graphing, or problem solving. Mathematics educators attributed the dismal performance of the majority of students in the nation's calculus classes to this traditional calculus curriculum. Consequently, by the mid-1990s, calculus reform movements had been initiated in many of the colleges and pre-college classrooms throughout the nation.

Calculus reform efforts at the college level in the 1990s often applied the pedagogical recommendations found in NCTM Standards. These pedagogical recommendations were also reflected in the revised Advanced Placement Calculus (AP Calculus) and International Baccalaureate Calculus (IB Calculus) courses offered in the nation's high schools. A measure of the subsequent success of the calculus reform movement at the pre-college level can be seen in the dramatic increase in numbers of students who took these courses from the 1980s into the twenty-first century. Specifically, the National Center for Education Statistics reported that the percentage of students completing calculus in high school had risen from 6% to 14% in the years from 1982 to 2004. The number of students completing calculus in high school continues to grow exponentially, at an estimated rate of 6.5% per year.

Several reform calculus curricula originated in the 1990s, and continue into the twenty-first century. The following examples are prominent reform calculus projects: Calculus, Concepts, Computers and Cooperative Learning (C^4L) conducted at Purdue University; the Calculus Consortium at Harvard (CCH) conducted at Harvard University; and Calculus and Mathematica (C&M) conducted at the University of Illinois at Urbana-Champaign and at Ohio State University.

While these three reform calculus projects differ from each other in significant ways, they share the following characteristics:

- They use graphing calculators, computers, and computer algebra systems (CAS) extensively for instruction, exploration, and visual representations. Supporters argue that technology serves to alleviate the huge burden of algebraic computation so characteristic of traditional calculus. The rationale for this reform is that technology facilitates instructional processes that focus on the principles of calculus rather than on computational procedures. Moreover, the graphical and visual representations provided by these technologies offer alternative modalities for learning that accommodate students' different learning styles. The curricula for CCH and C^4L focus heavily on graphing calculators, whereas the curriculum for C&M relies heavily on the computer software, Mathematica.
- The teacher serves as a facilitator of learning rather than the main conveyor of knowledge. While the teacher continues to initiate instruction and answer questions, mathematical situations are often explored by groups of students, using cooperative learning strategies. Using the principles of constructivist learning, students are guided to discover mathematical properties for themselves in a laboratory setting.
- A major focus is placed on real applications from multiple disciplines. The intention is to raise students' interest in the subject and motivate them with relevant applications.
- Mathematical rigor and formal language are de-emphasized. The abstractions of mathematical proof and rigor are postponed for several semesters to provide sufficient time for students to gain practical and intuitive knowledge of the subject.
- Assessment focuses heavily on students' writing, explanations of problem solutions, and open-ended projects. Sometimes students' explanations are valued as highly as the accuracy of their answers.

Whereas all of the above instructional practices have shown varying degrees of success in reform calculus classrooms, some areas of concern have been identified by those involved in the projects. Specifically:

- Focusing heavily on relevant applications sometimes results in the omission of important calculus content that cannot always be motivated by applications.
- The use of everyday language sometimes results in imprecise and incorrect mathematical definitions.
- Overuse of technologies for computation and graphing can weaken the development of students' quantitative reasoning and computational skills in calculus.
- Real-world problems are sometimes too complex and frustrating to students because of the extraneous and irrelevant information they usually contain.
- Short-answer problems for assessment are often easier for students than describing their problem-solving procedures in writing.

- Constructivist approaches are often too time consuming, allowing insufficient time for covering the entire calculus curriculum during class time.

Resolution of these concerns will surely be addressed in future curriculum revisions, and changes or modifications will be made accordingly. However, these accommodations are consistent with the historical evolution of calculus, which is the study of change and systems in perpetual motion.

Summary

In the past, calculus was taught in ways that made it accessible to only a small proportion of the population. However, recent curricular and pedagogical reforms in calculus, both at the college and pre-college levels, have served to increase student success, include twenty-first-century-technologies, and triple course enrollments. Statistics indicate that calculus enrollments will continue to increase exponentially. These findings suggest that calculus instruction in the United States is responding positively to the academic needs of society.

Indeed, by combining the power of technology with calculus, new areas of mathematics are emerging (for example, fractals, dynamical systems, and chaos theory). These new branches of mathematics have allowed humans to mimic nature's designs of mountain ranges, oceans, and plant growth patterns—which once were considered random acts of nature. In conclusion, calculus as a subject is still growing, and its applications are continually expanding to meet the needs of a dynamic, diverse, and technologically driven society.

Further Reading

Barnett, Raymond, Michael Ziegler, and Karl Byleen. *Calculus for Business, Economics, Life Science, and Social Science*. Upper Saddle River, NJ: Prentice-Hall, 2005.

Bressoud, David M. "AP Calculus: What We Know." June 2009. http://www.maa.org/columns/launchings/launchings_06_09.html#Q1.

Calinger, Ronald. *A Contextual History of Mathematics*. Upper Saddle River, NJ: Prentice-Hall, 1999.

Calter, Paul, and Michael Calter. *Technical Mathematics with Calculus*. 4th ed. Hoboken, NJ: Wiley, 2007.

Dubinsky, Edward. "Calculus, Concepts, Computers and Cooperative Learning." May 2004. http://www.pnc.edu/Faculty/kschwing/C4L.html.

Ferrini-Mundy, Joan, and K. Graham. "An Overview of the Calculus Curriculum Reform Effort: Issues for Learning, Teaching, and Curriculum Development." *The American Mathematical Monthly* 98, no. 7 (1991).

Gleaso, Andrew M., and Deborah H. Hallett. *The Calculus Consortium Based at Harvard University*. Spring 1992. http://www.wiley.com/college/cch/Newsletters/issue1.pdf.

International Baccalaureate Organization. *Diploma Programme Mathematics HL*. Wales, UK: Peterson House, 2006.

Johnson, K. "Harvard Calculus at Oklahoma State University." *The American Mathematical Monthly* 102, no. 9 (1995).

Murphy, Lisa. "Reviewing Reformed Calculus." http://ramanujan.math.trinity.edu/tumath/research/studpapers/s45.pdf.

Rogawski, Jon. *Single Variable Calculus*. New York: W. H. Freeman, 2008.

Silverberg, J. "Does Calculus Reform Work?" *MAA Notes* 49 (1999).

Steen, Lynn A. *On the Shoulders of Giants: New Approaches to Numeracy*. Washington, DC: National Academy Press, 1990.

Tucker, Thomas, ed. *Priming the Calculus Pump: Innovations and Resources*. Washington, DC: Mathematical Association of America, 1990.

Sharon Whitton

Cheerleading

Category: Games, Sport, and Recreation.
Fields of Study: Geometry; Number and Operations.
Summary: Cheerleading demonstrates and depends on an understanding of gravity and other forces.

Cheerleading is an activity that can be considered both recreation and a competitive sport, depending on the context. It typically consists of choreographed routines that require energy, discipline, and stamina, and may include chants, dance, tumbling, and other physical stunts. Cheerleaders make what they do look easy when, in reality, the underlying mathematics, such as symmetry, sequences, and physics, helps them to conquer gravity and fly. In 2008, the show *Time Warp* on the Discovery Channel analyzed the physics of cheerleading and gymnastics using slow-motion cameras.

History

In 1898, University of Minnesota football student Jack "Johnny" Campbell became the first person to lead football fans in cheers, using a megaphone, which had been invented by Thomas Edison in 1878, in order to spur his school's football team to victory. This cheering gave rise to organized cheerleading. Women joined the sport in the 1920s, bringing an opportunity to add gymnastics and throws to the cheerleading repertoire. Showmanship and pom poms were later added to the sport. The Dallas Cowboys cheerleaders' skimpy outfits in the 1970s changed the outward appearance of cheerleaders, while the 1980s brought the pursuit of more technical stunt sequences. In the new millennium, the *Bring It On* movies highlight the sport's challenges as well as its technical aspects. Although college squads are currently about 50% male, youth cheerleading is predominantly female. Cheerleaders are now found all around the world.

The Physics of Cheerleading

Cheerleaders are focused on center of mass and axes of rotation in order to maintain balance and complete pivots, jumps, and flips. Focusing on symmetry not only helps both their formations and individual poses have a more appealing look but also keeps them focused on maintaining an equal distribution of weight when they act as "bases" for a "climber" or "flyer. "

Cheerleaders need a firm grasp of gravity and the physics involved in their work, including Newton's Third Law, which states that for every action there is an equal and opposite reaction. For example, in a "full extension," the climber pushes off the two bases' shoulders and pulls up with his or her own shoulders to bear some the weight. The two bases move into a "chest prep" with their knees locked, their arms extended and locked, holding the climber's feet at chest level; the climber is now referred to as a flyer. The back person, or "spotter," will often be used as an additional holder to both hold some of the flyer's weight as well as to solidify the overall hold.

As the bases bend their knees, preparing to exert upward force in order to toss the flyer, each base's arms hold half the flyer's weight—uneven distribution of weight is seen when the bases' hips are uneven, exhibiting a loss of symmetry. The bases will extend their knees, letting go of the flyer's feet, to give the flyer upward force; the flyer lands exerting greater force on the way down, so the bases bend their knees and lock hands to cushion the catch. If the bases have not evenly distributed the weight, or have exerted unequal amounts of force, the flyer will not go straight up and the bases will need to move to catch the flyer.

In preparing to execute a flip, the cheerleader bends his or her knees to exert the upward force. To execute, the cheerleader needs to stay tight, keep the axis of rotation steady, point the feet, and land lightly, snapping together to a final pose to stop his or her momentum. The cheerleader's angular speed can change by changing the distance of mass to the axis of rotation; the cheerleader gets momentum from the push off as well as from

Focusing on symmetry helps to maintain an equal distribution of weight when acting as a base. (iStockphoto)

reducing the distance from mass to axis of rotation by tucking the body in as he or she rises from the ground.

Further Reading

Lesko, Nancy. "We're Leading America: The Changing Organization and Form of High School Cheerleading." *Theory and Research in Social Education* 16, no. 4 (1988).

Pennington, Bill. "As Cheerleaders Soar Higher, So Does the Danger." *New York Times* (March 31, 2007). http://www.nytimes.com/2007/03/31/sports/31cheerleader.html?_r=1&ref=cheerleaders.

Physics of Cheerleading. http://thephysicsofcheerleading.homestead.com.

Deborah L. Gochenaur

Climbing

Category: Games, Sport, and Recreation.
Fields of Study: Algebra; Data Analysis and Probability; Geometry; Problem Solving.
Summary: Effective climbing relies on mathematical principles, and there are connections between climbing and mathematical problem solving.

Climbing is the use of the human body and assisting equipment to ascend or descend steep surfaces. Climbing can be done professionally, such as for construction or in the military, for exercise or competition, or for performance—in the case of parkour. There are different styles of climbing depending on the object, such as bouldering, ice, tree, and rope climbing. If the weight of the climber is supported by equipment, it is called aid climbing; when the weight is supported only by the climber's muscles, it is called free climbing. Mathematics plays a role in successful climbing and in analyzing various aspects of the discipline. Mathematician Skip Garibaldi said, "Climbing has a lot of puzzles that have to be solved. It's not just strength or skill."

Anthropometry in Climbing

Anthropometry is the mathematical study of body measurements in order to understand human variability. For example, studies show that elite climbers, on average, tend to have small stature, low body mass, and a high handgrip-to-mass ratio compared to the population as a whole. Compared to nonclimber athletes with similar physical conditioning, they are frequently linear, with narrow shoulders relative to hips. Ape index is the ratio of a climber's arm span to height. In adults, it is usually close to one, as illustrated in Leonardo da Vinci's "Vitruvian Man." An ape index greater than one is reputedly advantageous for climbing, and some researchers have found ape index to be a statistically significant predictor of climbing success.

Fall Factor and Impact

Fall factor quantifies how hurtful a fall may be to a roped climber. Mathematicians such as Dan Curtis have derived the fall factor (F_{max}) using differential equations. It is a function of the ratio of the total distance the climber falls (D_T) to the length of the unstretched rope (L) between the climber and belayer or anchor at the rope's other end. It is also a function of the climber's mass (m), the elasticity or "stretchiness" of the rope (k), and gravity (g). Algebraically, it is represented as

$$F_{max} = \sqrt{2mgk\frac{D_T}{L}}.$$

Climbing ropes must pass a statistically designed drop test to be certified for sale and use. Other critical safety equipment is also designed using mathematics. One example is the curve of cams used in the "friend" devices that secure ropes to crevices in rock walls, which may be optimized using systems differential equations, sometimes with polar coordinates. The devices themselves are an application of logarithmic spirals.

Climbing Theories and Modeling

Many people have drawn parallels between climbing mountains and solving mathematical problems, especially great challenges like summiting Mount Everest and solving a problem like the Riemann hypothesis, first proposed by mathematician Bernhard Riemann. Analyses have shown that Everest climbers engage in multistep problem solving with altitude changes, rates, percentages, conversions, approximations, and division of large numbers. Mathematician-climber John Gill said that problems in both mathematics and climbing are often solved by "quantum jumps of intuition." Patterns found in the natural features of some popular climbing locations can very mathematical. The

Navajo Sandstone formation includes rounded domes and saddle shapes with remarkably precise-looking contour lines.

At the same time, the geometric diversity and complexities of climbing surfaces and the variety of techniques used by climbers have made developing a single theory of optimal climbing strategy difficult. However, several methods are used to quantify characteristics of different climbs and probabilistic models can be used to make decisions. Competitive climbers assign climbing grades to climbing routes, using objective and subjective criteria, to describe their difficulty. Other systems assess the technical difficulty of required moves, the stamina necessary, exposure to the elements, or the frequency of difficult moves. Mathematician Alan Tucker demonstrated using graph theory that the classic Parallel Climbers mathematical puzzle has a solution for any mountain range.

Further Reading

Curtis, Dan. "Taking a Whipper: The Fall-Factor Concept in Rock-Climbing." *The College Mathematics Journal* 36, no. 2 (2005).

Garlick, S. *Flakes, Jugs, and Splitters: A Rock Climber's Guide to Geology.* Kingwood, TX: Falcon, 2009.

Tucker, Alan. "The Parallel Climbers Puzzle." *Math Horizons* 3 (November 1995).

Maria Droujkova

Cocktail Party Problem

Category: Friendship, Romance, and Religion.
Fields of Study: Data Analysis and Probability; Measurement.
Summary: A metaphorical cocktail party is the setting for a source separation problem and other challenges.

The eponymous Cocktail Party Problem is a source separation problem in digital signal processing, wherein digital systems have difficulty separating out one signal among many—the metaphorical conversation in a noisy cocktail party, which is comparatively easily handled by the human brain. More broadly, distinguishing signal from noise is a data analysis challenge with many specific applications. The metaphor of the cocktail party also lends itself to a number of other problems in combinatorics, graph theory, probability, and functional analysis.

Conversations and Background Noise

With all the noise at a party, it can be difficult to focus on one conversation, although many people are able to do so. Telecommunication professor Colin Cherry conducted experiments in this area, and he is considered by some to be a pioneer in cognitive science. Many people can even recognize the sound of their name from across a noisy room. However, this is not as easy when heard on a recording. One cocktail party problem arises from concerns about separating each individual's voice characteristics in a recording from the other voices and background noise. People in surveillance and intelligence are inherently interested in such a problem, and scientists and engineers have worked on solutions since at least the 1950s. One common method is mathematical signal processing. Mathematicians and engineers digitize a signal using a Fourier transform, named for Joseph Fourier. They process it using a variety of methods to remove noise and other extraneous information, and then reconstruct the signal using the inverse transform.

While the process may result in an improved recording of one person's voice, early twenty-first-century technology and methods do not provide perfect separation, so the recording still includes at least some distracting background noise. However, engineers have conjectured that the signal should be able to be reconstructed without the noisy phase. Mathematicians Radu Balan, Peter Casazza, and Dan Edidin made progress on the problem in 2006, when they showed—using a neural net—that it is mathematically possible to retain the voice characteristics without the noise. Scientists continue to work on developing algorithms for practical use. Casazza made another fundamental mathematical discovery during his work on the cocktail party problem. He and his wife, Janet Tremain, also a mathematician, showed that the Kadison–Singer problem, named for mathematicians Richard Kadison and Isadore "Iz" Singer, is equivalent to other unsolved problems in areas of pure and applied mathematics and engineering, such as operator theory, harmonic analysis, and signal processing.

Mathematicians also investigate other party problems, like the probability that when people at a party are chosen to be partners for a card game—like bridge—no randomly chosen partners will contain spouses or members of the same family. The solutions require finding specific combinations or permutations of the guests. Under certain constraints, the maximal probability for some problems may be bounded at less than certainty as the number of people at the party grows. There are also connections between this question and the card game War, as well as with a related set of problems that focus on orders and arrangements of guests around a single dinner table or in various groupings, with applications in areas like queuing theory and assignment problems. The classic dining philosophers problem is yet another variation that has applications in resource sharing and task allocation in computer science.

Another party problem asks how many people must be present at a party in order to ensure that there will be a group of three people who share the characteristic of being acquaintances or strangers. There is no guarantee that three people will all know each other or will be strangers in parties of five or less people since counterexamples exist. The Java game HEXI, named so because the game is played on the vertices of a hexagon, is modeled on this question. The six vertices are connected by edges and each player takes a turn coloring an edge his or her color. One color represents acquaintances, and the other represents strangers. The goal of the game is to avoid making a triangle of the same color. Mathematicians model this question using graph theory, and show that in any group of at least six people, it is possible to find a group of three people satisfying one of the mutually exclusive relationships. Hence HEXI will always have a loser. Instead of people at a party or vertices of a polygon, one could explore other objects like nations embroiled in a conflict, sequences of randomly generated numbers, or stars. Mathematicians investigate problems like these concerning the existence of regular patterns in sets of objects in Ramsey theory, named for Frank Ramsey.

Further Reading
Albertson, Michael. "People Who Know People." *Mathematics Magazine* 67, no. 4 (1994).

Brodie, Marc. "Avoiding Your Spouse at a Party Leads to War." *Mathematics Magazine* 75, no. 3 (2002).
Casazza, Peter, and Janet Tremain. "The Kadison–Singer Problem in Mathematics and Engineering." *Proceedings of the National Academy of Sciences* 103, no. 7 (2006).
"Mathematicians Solve the 'Cocktail Party Problem.'" PHYSorg.com. August 22, 2006. http://www.physorg.com/news75477497.html.

SARAH J. GREENWALD
JILL E. THOMLEY

Composing

Category: Arts, Music, and Entertainment.
Fields of Study: Algebra; Number and Operations; Representations.
Summary: Mathematics and music developed in tandem and composition is firmly grounded in mathematics.

Throughout the history of Western music, composers have utilized mathematical techniques in creating musical works. From Pythagoras, Plato, and Ptolemy in ancient Greece to the sixth-century music theorist Boethius, music was thought to be a corollary of arithmetic. With the widespread development of modern standardized musical notation thought to have begun in the Renaissance, compositional craft became more highly developed. Compositions intertwined with mathematical patterns were particularly highly regarded.

The eighteenth-century composer and theorist Jean-Philippe Rameau was unequivocal in his views on the connection between mathematics and music in his 1722 *Treatise on Harmony*, writing, "Music is a science which should have definite rules; these rules should be drawn from an evident principle; and this principle cannot really be known to us without the aid of mathematics." Fugal composition techniques in the high Baroque period were highly mathematical. The classical and romantic eras, characterized by a movement away from polyphonic music, produced less obvious mathematically oriented composition technique. In the twentieth century, however, mathematical formalisms were fundamental as replacements for the tonal structures of the romantic era. There are even subgenres of rock music (started in the 1980s) called "math rock" and "mathcore" (after metalcore, a fusion of heavy metal and hardcore punk), which uses complex and atypical rhythmic structures, angular melodies, unusual time signatures, and changing meters. Metalcore, in particular, also uses harmonic dissonance. In another example, Robert Schneider composed a mathematical score for a play in 2009. He said:

> I wrote a composition called 'Reverie in Prime Time Signatures,' that is obviously written in prime time signatures, that is, only prime numbers of beats per measure. Also the piece has kind of a sophisticated middle section that encodes some ancient Greek mathematics related to prime numbers in musical form, that I am proud of.

The Renaissance Canon

During the Renaissance, mathematical devices were developed to a considerable degree by Northern European composers. In the canons of Johannes Ockeghem, a single melodic voice provides the basis by which one or more additional voices are composed according to various mathematical transformations of the original: mirror reflection of musical intervals (inversion), time translation, mirror reflection in time (retrograde), or a non-unit time scaling (mensuration canon). Composers of this period understood the word "canon" to mean a rule by which secondary voices could be derived from a given melody, in contrast to our modern usage of the word, which means a simple duplication with later onset time, as in the nursery rhyme round "Row, Row, Row Your Boat."

Mathematical Transformations in Composition

In addition to standard musical notation, music can be represented mathematically as a sequence of points in an algebraic structure. A musical composition can be represented as a sequence of points from the module M over the cyclic groups of integers Z_p

$$M = Z_{p_1} \times Z_{p_2} \times Z_{p_3} \times Z_{p_4},$$

with the coordinates representing (respectively) onset time, pitch, duration, and loudness. For example, the 12 notes of the chromatic scale would be represented in the second coordinate by Z_{12}. In this schematic, if a point (x_1, x_2, x_3, x_4) in a musical motif were repeated later at a different volume level, the repetition would differ in the first and last coordinate and would be represented as $(x + \alpha, x_1, x_2, x_3, x_4 + \beta)$, where α is the time shift and β is the amount of the volume difference.

Inversion takes the form $(x_1, 2\alpha - x_2, x_3, x_4)$. Mensuration, as in the canons of Ockeghem, is written $(x_1, x_2, \alpha \cdot x_3, x_4)$. Transformations of this form were used extensively in the Renaissance and Baroque eras and played a fundamental role in post-tonal era of the twentieth century.

Mathematical Structure in Atonal Music

At the turn of the twentieth century, music theorists and composers looked for new organizing principles

Bach: The Canon Master

Johann Sebastian Bach was a master of canonic composition. Bach's canons challenged performers to solve puzzles he set before them. Examples abound in A Musical Offering (BWV 1079), written in 1747. The first of two Canon a 2 (canon for two voices) from Musical Offering appears to have two different clef symbols: one at the beginning of the first measure, and one at the end of the last. The first singer had to read from beginning to end, and the second had to start at the same time and read in the opposite direction. In this small piece, Bach provides an example of retrograde or cancrizan (crab) canon. The puzzle in the second Canon a 2 is even more cleverly concealed: a single line with two clef signs in the first measure, one upside down. The cryptic instruction Quaerendo invenietis ("Seek and ye shall find") is inscribed at the top of the manuscript.

The second, inverted clef sign indicates that the second voice of the canon is to proceed in inversion, and the performer is left to "seek" the appropriate time translation at which the second voice should begin. Another example of Bach's masterful canonic treatment is BWV 1074: Kanon zu vier Stimmen, which with its numerous key signatures, clefs, and repeat signs can be played from any viewing angle.

on which atonal music could be structured. Groundbreaking composer Arnold Schoenberg turned to the idea of "serialism," in which a given permutation of the 12 chromatic pitches constitutes the basis for a composition. The new organizing principle called for the 12 pitches of this "tone row" to be used—singly, or as chords, at the discretion of the composer, always in the order specified by the row. When the notes of the row have been used, the process repeats from the beginning of the row.

Composers like Anton Webern, Pierre Boulez, and Karlheinz Stockhausen consciously used geometric transformations of onset time, pitch, duration, and loudness as mechanisms for applying the tone row in compositions. In the latter half of the twentieth century, set theoretic methods on "pitch class sets" dominated the theoretical discussion.

Predicated on the notions of octave equivalence and the equally tempered scale, Howard Hanson and Allen Forte developed mathematical analysis tools that brought a sense of theoretical cohesion to seemingly intractable modern compositions. Another mathematical approach to composition without tonality is known as aleatoric music, or chance music. This technique encompasses a wide range of spontaneous influences in both composition and performance. One notable exploration of aleatoric music can be seen in the stochastic compositions of Iannis Xenakis from the 1950s. Xenakis's stochastic composition technique, in which musical scores are produced by following various probability models, was realized in the orchestral works *Metastasis* and *Pithoprakta*, which were subsequently performed as ballet music in a work by George Ballanchine.

Further Reading

Beran, Jan. *Statistics in Musicology*. Boca Raton, FL: Chapman & Hall/CRC Press, 2003.

Forte, Allen. *The Structure of Atonal Music*. New Haven, CT: Yale University Press, 1973.

Grout, Donald Jay. *A History of Western Music*. New York: Norton, 1980.

Temperley, David. *Music and Probability*. Cambridge, MA: MIT Press, 2010.

Xenakis, Iannis. *Formalized Music: Thought and Mathematics in Composition*. Hillsdale, NY: Pendragon Press, 1992.

Eric Barth

Connections in Society

Category: School and Society.
Fields of Study: Connections.
Summary: An integrated approach to mathematics stresses the importance of making connections among various perspectives and applications.

While mathematics in educational settings is often separated out into differing subjects, it is important to understand that mathematics is an interconnected field of study. While most individuals are aware that they must be familiar with basic addition and subtraction to ensure the proper handling of money, very few individuals give any thought to the multitude of deeper mathematical connections they experience daily. In fact, both the National Science Foundation and the National Council of Teachers of Mathematics have recently begun to strongly advocate for the use of an interconnected curriculum in K–12 mathematics education. An integrated approach to mathematics education stresses the importance of making connections among mathematical perspectives, as in algebra and geometry, making connections to other fields, as in physics or religion, and connecting mathematical concepts to society as a whole, as in applications and usefulness in daily living.

The purpose of an interconnected curriculum is to help students better understand how the various branches of mathematics are connected and how mathematics is connected to the real world. By teaching mathematics as a unified whole, rather than multiple discrete subjects, students may better understand that mathematics is not a set of indiscriminate rules and isolated skills; rather, it involves a rich interplay between mathematical concepts, as well as complex interactions with other academic subjects. It is this integrated approach to mathematics that seeks to answer that question, "When are we ever going to use this in *real life?*" When this objective is met, students often show an increased appreciation and enthusiasm for mathematical principles.

People use many different interrelated approaches to process ideas, analyze objects, make decisions, or solve problems. For example, one might calculate the optimal viewing distance of a painting in order to see the depth that the artist intended, examine the surface of the painting to appreciate the finer details and glazes, or stand back to appreciate the overall effect and balance of colors. Real-life situations are not divided the way they are in textbooks by their applicability to a certain topic or technique, like exponential models. In fact, throughout the twentieth century, employers, such as engineering firms, complained about the lack of connections made in school between different subjects. Mathematician Eliakim Moore discussed this problem a century ago in his 1902 address as the president of the American Mathematical Society. In 1989, the National Council of Teachers of Mathematics published a set of national standards for mathematics that included connections as a significant component.

Whereas traditional mathematics curricula in the twentieth century separated subject areas like algebra and geometry, an integrated approach involves presenting mathematical subjects as one interrelated whole that also connects to other subjects and real-world experiences. In antiquity, the square of a number was defined as the area of a square with the same side length. People with interdisciplinary interests were perhaps more common at that time—Greek mathematicians were also astronomers, inventors, engineers, and philosophers. Throughout history, mathematicians such as Carl Friedrich Gauss contributed to so many areas of mathematics and to other fields, like geodesy; but in the twenty-first century, researchers who specialize in a subdiscipline are more common. However, connections among multiple mathematical perspectives are still important in the development of mathematics. Algebra and geometry remain linked and the field of algebraic geometry is active today. Many researchers use techniques from a variety of different mathematical fields. Geometers may heavily rely on concepts from analysis, linear algebra, number theory, or statistics, for example. Other researchers work in the intersection of fields like statistical analysis.

Mathematics can easily be connected to other scientific disciplines, like physics or biology. Mathematics is sometimes referred to as the "foundation" or "language" of science. However, there are many other types of links between mathematics and the sciences. Some researchers work on problems at the interface of mathematics and a scientific field, while others translate ideas from the sciences to solve problems in mathematics and vice versa. Scientific disciplines like physics are often

Mathematics and the FIFA World Cup Finals

Mathematics is also used in one of the most popular sporting events in the world: the FIFA World Cup Finals. Thirty-two teams qualify for the World Cup Finals, and they are assigned to eight groups of four teams. The top seven teams in the world and the host country's team are seeded based on the FIFA World Rankings and recent World Cup performances and put into the eight different groups. The rest of the teams are put into different pots based on their geographical location, and then teams from each pot are randomly assigned to the eight groups. In the group stage of the World Cup Finals, each team plays every other team in its group.

A team earns three points for a win, one point for a tie, and zero points for a loss. In each group, the two teams with the most points advance to the knockout stage. If teams are equal in points, the greatest goal difference, the greatest number of goals scored, and other various statistics can be used to determine the top two teams to advance to the next stage. Sixteen teams advance to the knockout stage, which is a single-elimination tournament. At the end of the tournament, FIFA crowns one World Cup champion, as well as several individual awards, such as the Golden Ball for the best player and the Golden Boot for the top goal scorer. The winner of the Golden Ball award is based on a vote of media members. The Golden Boot award is given to the player with the greatest number of goals scored, as well as with the greatest number of assists. Mathematics is used in important calculations of sports statistics, college and world rankings, tournament rankings, and awards for individual performances.

referred to as partner disciplines for mathematics. Researchers have met for a conference named "Connections in Geometry and Physics" that explores the interdisciplinary facets. In geometry and physics there is a concept called a "connection," which is an operator that allows for comparison at different points in a space via parallel transport. Mathematics has been interwoven with physics since antiquity. There have also been historical linkages between mathematics and biology, but the interdisciplinary field of mathematical biology has grown rapidly in the early twenty-first century.

Figure 1. A Common Nutrition Facts Label.

Nutrition Facts
Serving Size 1/2 cup (about 82g)
Servings Per Container 8

Amount Per Serving	
Calories 200	Calories from Fat 130
	% Daily Value*
Total Fat 14g	**22%**
Saturated Fat 9g	**45%**
Trans Fat 0g	
Cholesterol 55mg	**18%**
Sodium 40mg	**2%**
Total Carbohydrate 17g	**6%**
Dietary Fiber 1g	**4%**
Sugars 14g	
Protein 3g	
Vitamin A 10%	Vitamin C 0%
Calcium 10%	Iron 6%

*Percent Daily Values are based on a 2,000 calorie diet. Your daily values may be higher or lower depending on your calorie needs:

	Calories:	2,000	2,500
Total Fat	Less than	65g	80g
Saturated Fat	Less than	20g	25g
Cholesterol	Less than	300mg	300 mg
Sodium	Less than	2,400mg	2,400mg
Total Carbohydrate		300g	375g
Dietary Fiber		25g	30g

Calories per gram:
Fat 9 • Carbohydrate 4 • Protein 4

Students may have difficulty appreciating the importance of mathematics in nonscientific disciplines, but the connections between mathematics and subjects like business, art, music, or religion are multilayered and multifaceted. For example, mathematics has played a part in religious life since the earliest documented cultures. The ancient Mesopotamians, embracing a polytheistic faith, developed the time system we use today with bases of 60 (60 seconds make up a minute, and 60 minutes comprise an hour). Adherents of Christianity, Judaism, and Islam have all embraced elements of mathematics in the conceptualization of sacred time. Given the importance of religion today, this time is still of great value for humankind. Mathematics plays a key role in the calculation of religious celebrations around which many faiths flourish. The week and solar day provide a delineation of sacred days that are different from the others—Sunday for Christians, Saturday for Jews, and Friday for Muslims. In other ways, numeric or geometric symbolism plays a significant part of religious practice.

There are countless examples that highlight the importance of mathematics in daily tasks. In the twenty-first century, it is almost impossible to find a task that does not connect to mathematics, either directly or through the tools and technologies in which mathematics plays an important role. In turn, mathematicians formulate new theories and concepts in order to meet the needs of society.

Mathematics as a Universal Language

Many people consider mathematics as the only truly universal language, regardless of gender, culture, or religion. For example, while the precise number of digits that are used in applications may differ, the ratio of the circumference to the diameter of a circle is still π, irrespective of the cultural context. Calculating the cost of groceries involves the same mathematical processes whether one is paying for those groceries in dollars, pesos, or pounds. With the universal language of mathematics, regardless of the unit of exchange, humans are likely to arrive at similar mathematical results. In fact, there are many examples of researchers in different areas of the world who independently arrived at the same theorems. Thus, mathematics as a universal language provides a common ground, creating the capacity for human beings to connect to one another across continents and across time.

Nutrition Labeling

An important way that mathematics can be found in our everyday life is on nutrition facts panels, which are mandated by the Nutrition Labeling and Education Act of 1990 to be placed on nearly all multiple-ingredient foods. The nutrition facts label on foods must list the fat, saturated fat, trans fat, cholesterol, sodium, total carbohydrate, fiber, sugar, protein, Vitamin A, Vitamin C, calcium, and iron content of the food. Other nutrients may be listed voluntarily. These labels also include a column that lists the percent Daily Value (% DV) to help consumers decide whether the nutrient content of a serving of the food product is a lot or a little. Mathematics is used to calculate the calories per serving and the % DV of a serving listed on the nutrition facts label.

As shown in Figure 1, at the top of the nutrition facts label, the serving size, as well as the number of servings per container, is listed directly underneath "Nutrition Facts." In this case, a serving size is ½ cup and there are eight servings per container. This means that there are four cups (½ cup × eight servings = four cups) of food in this package. If a person consumed half the container, or two cups of food, he or she would have had four servings (the amount of food consumed divided by a serving size, or two cups divided by ½ cup per serving = four servings).

Next, the calories per serving and the calories from fat per serving are listed. In this food, there are 200 calories per serving and 130 calories from fat in one serving. If the person consumed four servings and there are 200 calories per serving, then he or she consumed 800 calories (four servings × 200 calories/serving = 800 calories). Similarly, this person consumed 520 calories from fat (four servings × 130 calories from fat/serving = 520 calories from fat).

Following the calorie content, the nutrition facts label also lists the number of grams of total fat, total carbohydrate, and protein, which are calorie-yielding nutrients. A gram of fat contains nine calories, which is listed at the very bottom of the label. In this food, a single serving contains 14 grams of fat, which yields 126 calories (14 grams of fat × 9 calories/gram of fat = 126 calories from fat). This calculation was done to create the number of calories from fat listed on the panel (they rounded up to 130). As previously mentioned, if a person ate four servings, he or she consumed about 520 calories from fat.

The number of calories from carbohydrates and proteins can also be calculated. Both carbohydrates and protein yield four calories per gram, which is also listed at the very bottom of the nutrition label. In this food, there are 17 grams of carbohydrates, which provides 68 calories (17 grams × 4 calories/gram = 68 calories). In four servings, a person would ingest about 272 calories from carbohydrates (68 calories/serving × 4 servings = 272 calories from carbohydrates). There are three grams of protein in one serving, which means there are 12 calories from protein in one serving (3 grams × 4 calories/gram = 12 calories) and 48 calories from protein in four servings (12 calories/serving × four servings = 48 calories).

On the right side of the nutrition facts panel, the % DV is also listed. These daily values are based on a 2000-calorie diet, which is stated on the label next to the asterisk. Near the bottom of the label, it lists the maximum number of grams or milligrams of total fat, saturated fat, cholesterol, or sodium that a person should consume per day if on a 2000-calorie diet. It also lists the number of grams of total carbohydrate and fiber a person should eat if on a 2000-calorie diet.

If there are 14 grams of fat in one serving of this food and a person on 2000-calorie diet should consume no more than 65 grams of fat per day, then one serving of this food yields 22% of a person's DV of fat (14 grams of fat/65 grams of fat = about 22%). If this person has consumed four servings, then he or she has eaten 88% of his or her DV of fat (22%/serving × four servings = 88%). The same calculations can be made for the saturated fat, cholesterol, sodium, total carbohydrate, and fiber. Similar calculations are also made for the vitamins listed on a nutrition facts panel.

As demonstrated, mathematics is used in the calculations surrounding calorie content and % DV on nutrition labels. The mathematics used can affect a person's choice of foods and, in turn, a person's health.

Sports

Mathematics is used in numerous other everyday activities, such as sports. It is common in popular sports to calculate statistics to measure performance. In baseball, a common statistic is a batting average. A batting average is a simple calculation: the number of "hits" divided by the number of "at bats." This statistic is used to estimate an individual's batting skills. In professional baseball, a batting average of .300 is considered an excellent batting average.

A similar statistic to the batting average is in volleyball, which is called a hitting percentage. However, it is slightly different because it tries to measure an individual's hitting or attacking skills and takes errors into accounts. It is calculated by taking the number of kills, subtracting the number of errors, and then dividing the difference by the number of attempts. A "kill" is when a hitter's attack results directly in a point (the ball falling into the opponent's area of the court, an opponent not being able to return the ball, or the opponent making a blocking error as a result of the attack). An "error" is when a player hits the ball and it goes into the net (does not cross to the opponent's side) or out of bounds. An "attempt" is anytime the player tries to attack the ball. For example, if a player had 10 kills, 3 errors, and 17 attempts, the player's hitting percentage would be about .412 ($(10-3)/17 = 0.412$), which would also be considered a good hitting percentage, similar to the guidelines to the batting average.

Mathematics is important in the calculation of college football Bowl Championship Series (BCS) rankings as well. A mathematical formula is used to calculate these rankings, which order the top 25 NCAA Division I-A football teams based on their performance during the prior week. At the end of the season, the top two teams play each other in the national championship bowl. Mathematical formulas are also used to calculate which teams will play in the other bowls, taking into consideration the conference the team comes from and how many fans and advertising dollars the team is likely to bring in as well.

More specifically, the main factors that go into these rankings are subjective polls, computer rankings, the difficulty of a team's schedule, and the number of losses. The subjective poll numbers come from the average of two rankings from the Associated Press (AP) and the *USA Today*/ESPN Coaches Poll Ratings. Sports writers and broadcasters vote in the AP poll and a select group of football coaches vote in the *USA Today*/ESPN Coaches poll on which football teams they think are the best, and then these two rankings are averaged. The computer rankings are based on eight different computer rankings that are calculated based on a team's statistics for that week (strength of the opponent, final score, win-loss record, and so forth). The strength of a team's schedule is based on a cumulative win-loss record of its opponents, as well as their opponent's opponents. The calculation of the number of losses is straightforward. Each loss that a team suffers corresponds to one point, which is added to its final score. Points from each category are assigned to the team, and then these values are added to create a team's final score. The team with the lowest point total is ranked "number one" in the rankings.

Speedometers

Mathematics is also used in cars. All cars have a speedometer, which is a device used to calculate an instantaneous speed of a vehicle. It is important for a driver to know the speed of the vehicle at all times to ensure the safety of passengers and pedestrians and to abide by local traffic laws. In the United States, speedometers are read in terms of miles per hour. The calculation of the speed of the vehicle requires significant mathematics.

In many vehicles, an eddy current or mechanical speedometer is used, which is the speedometer with a needle that points to the speed that the vehicle is travelling. In these cars, there is a drive cable that runs from the speedometer to the transmission, which has a gear that tracks the rotational speed of the wheels. In other words, the gear tracks the number of revolutions the wheel makes within a certain time frame. Digital speedometers calculate miles per hour slightly differently, using a vehicle speed sensor. The vehicle speed sensor is in the transmission and also tracks the rotations of the wheels. From this information, the vehicle's speed is calculated and displayed on either a digital screen or a traditional needle-and-dial display.

The calculation of a vehicle's speed is dependent on the size of the tire as well. For example, if the tire rotates x times per minute, then the vehicle's speed can be calculated in miles per hour. Knowing the diameter of the tire, the circumference of the tire can be calculated (diameter $\times \pi$). Therefore, the vehicle travels the distance of the number of revolutions times the circumference of the tire, within a certain time frame. This ratio can then be converted to miles per hour by converting the units. Because all of these calculations are based on an assumed tire diameter and circumference, it is very important for drivers to ensure that the correct size tires are on their vehicle. If a car's wheels are too large or too small, the speedometer will read slower or faster than the vehicle's actual speed, which may lead to accidents, speeding tickets, or just slower driving.

Conclusion

Mathematics can be found in everyday situations that have a real and important effect on our lives. All areas of one's life are in some way connected to mathematical principles. Only a small number of examples have been presented here—the list can be expanded infinitely. In fact, one would be hard pressed, in today's technologically advanced world, to present even a handful of activities that do not involve some mathematical concepts, if even at the unconscious level. By bridging the disconnect between "school mathematics" and "real-life mathematics," individuals gain a greater appreciation for—and curiosity of—mathematical applications.

By viewing mathematics as an integrated whole and understanding its connectedness to society, individuals become active participants, rather than passive recipients, of information. When one becomes aware of mathematical connectedness, rather than viewing math as a series of isolated and disconnected concepts to be learned though rote memorization, an individual develops the understanding of mathematics as a crucial and meaningful tool that can aid in the understanding, predicting, and quantifying of the world around us.

Further Reading

Brookhart, Clint. *Go Figure: Using Math to Answer Everyday Imponderables.* Chicago: Contemporary Books, 1998.

Cuoco, Al. *Mathematical Connections.* Washington, DC: The Mathematical Association of America, 2005.

Garland, Trudi H., and Charity V. Kahn. *Math and Music: Harmonious Connections.* Palo Alto, CA: Dale Seymour Publications, 1995.

House, Peggy, and Arthur Coxford. *Connecting Mathematics Across the Curriculum.* Restin, VA: National Council of Teachers of Mathematics, 1995.

Martin, Hope. *Making Math Connections: Using Real World Applications With Middle School Students.* Thousand Oaks, CA: Sage, 2007.

Lee Anne Flagg
Matthew West
Kristi L. Stringer
Casey Borch

Cooking

Category: Arts, Music, and Entertainment.
Fields of Study: Number and Operations; Measurement.
Summary: A good cook must be able to compute conversions, costs, and measurements.

In his Renaissance play, *The Staple of News,* Ben Jonson likens a master cook to—among other things—a mathematician. Although many people would think this comparison is an exaggeration, the mathematical requirements placed on the modern cook are significant.

In the past, cooking skills were passed on orally and through apprenticeship from generation to generation; today, inexperienced cooks are expected to learn to cook from recipes, which consist of a list of measured ingredients followed by instructions that refer to temperatures, times, and possibly more esoteric measurements. In addition to being able to scale recipes, the cook in our global world encounters many interesting recipes from diverse cultural traditions, which use a variety of systems of measurement. Cooks must also be able to plan healthy and cost-effective menus.

Measurement of Ingredients

In recipes written in the United States, quantities for both liquid and dry ingredients are often specified by volume, and are measured in terms of teaspoons, tablespoons, or cups, in which there are 3 teaspoons to 1 tablespoon, 16 tablespoons to 1 cup, and 2 cups to 1 pint. Special measuring cups are made that permit the leveling of dry ingredients to ensure precise measurement. For measuring liquid ingredients, different cups are used that have graduation marks down the side and a convenient pouring spout. Measuring spoons are used for smaller quantities of both liquid and dry ingredients. For an experienced cook, the quantities given in recipes serve as general indications; however, in baking, when certain chemical reactions are expected to be balanced, precision is needed.

For more consistent outcomes, quantities are specified by weight. Ingredient densities vary. For example, a cup of water weighs 8 ounces, whereas a cup of flour—depending on how it was scooped—weighs about 5 ounces. Tables to assist in conversion between weight and volume can be found on the Internet. There can be confusion with the word "ounces," which can refer to either weight or volume. Ounces used for dry ingredients refer to one-sixteenth of a pound. Ounces measuring liquid ingredients refer to either one-sixteenth of a pint or to one-twentieth of a pint, depending on what is being measured.

Modern recipes written outside the United States provide measurements in the metric system. Liquid ingredients are specified in liters (volume) while dry ingredients are specified in grams (mass). Since kitchen scales actually measure weight, most cooks view grams as measuring weight. One liter of water weighs approximately 1000 grams. A liter is 1000 cubic centimeters, or about 1.057 quarts. A kilogram, 1000 grams, is approximately 2.205 pounds. A deciliter is one-tenth of a liter and is often used for recipes designed for home use. The metric system—based on multiples of 10—is designed to simplify calculations and scaling of measurements and is becoming the preferred system for cooks.

Scaling a Recipe

Recipes often specify the number of portions that they produce. To alter the number of portions generated, the recipe is scaled. This involves multiplying the quantity of each ingredient by a scale factor. To double a recipe, the scale factor is 2, while to halve a recipe, the scale factor is 1/2. At times, a more complex scaling is required. For example, imagine a baker is following a recipe that calls for 125 grams of pre-fermented dough. The recipe to make pre-fermented dough calls for 1000 grams of flour, 10 grams of yeast, and 0.6 liters of water and results in 1610 grams of dough. Since only 125 grams of pre-fermented dough are needed, the required scale factor is 125/1610 = 0.078.

A naïve scaling results in 78 grams of flour, 47 grams of water (.047 liters), and the absurdly small amount (0.78 grams) of yeast. An experienced cook would add more yeast. Most recipes written for home use can only be scaled up or down by less than a factor of 4. Additionally, some ingredients, like spices, gelatin, and leavening agents, should not be scaled proportionately. Most good general cookbooks will give advice on scaling recipes. A good collection of professional recipes for large numbers of portions is available from the Armed Forces Recipe Service.

Measuring Temperature

Controlling temperatures on most modern stovetops is easier than doing so on wood-burning stoves. However, techniques vary significantly among gas, electric, and induction cookers and are best described by the manufacturer. In some instances, such as deep fat frying or candy making, temperature on the stovetop is

Molecular gastronomy is a new trend in cooking with a scientific slant. A chef plates a dish called strawberry ravioli created using reverse spherification, and places so-called caviar spheres of sauce with chop-sticks. (iStockphoto)

measured by a thermometer. In making candy sugar syrup, temperature can also be measured by "feel" or by the way a drop of the syrup interacts with cold water. Books on making candy describe the relationships among these methods. The temperature of an oven is accurately monitored by a thermostat, which can be set. Often, an oven thermometer is also used to check the oven thermostat. Most recipes give the required temperature in either Fahrenheit or Celsius (previously called centigrade). The formula for converting from Fahrenheit to Celsius is given by

$$C = \frac{5}{9}(F - 32)$$

and from Celsius to Fahrenheit by

$$F = \frac{9}{5}(C + 32).$$

Thus, an oven temperature of 350 degrees Fahrenheit is about 177 degrees Celsius. Temperatures in some older British recipes are given in gas mark settings, in some older French recipes in numbered settings, in some older German recipes as Stufe settings, and in some much older recipes as verbal descriptions such as Very Slow or *Doux*. Tables showing conversions among these various approaches to measuring temperature can be found in general cookbooks and on the Internet.

Other Important Measurements

Other important quantities that need to be measured when cooking include time, acidity, and density. Time measured in seconds, minutes, and hours—a system based on 60—is now probably universal. Because estimating the passage of time is fraught with error, early recipes specified important times "as measured by the clock." Acidity is measured on the pH scale. Water, which is neutral, has a pH of 7. An acidic solution, like orange juice, might have a pH of 3, while a basic solution of baking soda in water might have a pH of 9. In home cheese making, the conversion of lactose to lactic acid is tracked by monitoring pH levels of the milk; however, traditional cheese makers will use the Dornic scale.

Measuring the density of a solution is important in wine and beer making, and in candying fruits. For example, the density of fresh grape juice indicates the ripeness of the grapes and the alcohol content of the finished wine. Candying fruit in sugar water can take many days. The daily gradual increase of sugar in the syrup where the fruit is steeping maximizes the amount of sugar absorbed by the fruit. The density of the syrup is carefully checked to ensure the correct increase of sweetness. Density of syrups is measured with a hydrometer, and a variety of scales, including Brix, Baumé, and specific gravity, have been used in recipes. Although older French recipes will refer to the Baumé scale, since the 1960s, most recipes have used specific gravity. For syrups that are denser than water, a simple approximate conversion from Baumé to specific gravity (*sg*) is given by:

$$sg = \frac{145}{145 - °B}.$$

Menu Planning and Budgeting

Cost and nutrition are also important factors for cooks. Many modern recipes, in addition to giving calories per serving, will give grams of carbohydrates, protein, fat, cholesterol, sodium, and calcium. This information, along with labels on prepared food, helps guide the cook in making nutritional choices. A cook might also be interested in knowing the cost of a portion size. For example, consider a portion of boneless chicken breast. The cost as purchased is what the chicken breast with bone costs per pound. Once the breast has been boned, what remains weighs less and results in a higher cost per pound of the edible portion. During cooking, the breast will shrink, resulting in an even higher cost per pound of the breast as served. Being aware of these costs, along with labor costs and inventory costs, helps the cook determine the cost of each item served. Although the home cook probably does not go through all these computations, a good home cook will have an idea of monthly food expenditures and how these costs are distributed among the various kinds of food served.

Further Reading

Bilheux, R., and Alain Escoffier. *Creams, Confections, and Finished Desserts*. Hoboken, NJ: Wiley, 1998.

Haines, R. G. *Math Principles for Food Service Occupations*. 3rd ed. Albany, NY: Delmar Publishers, 1996.

Jones, T. *Culinary Calculations: Simplified Math for Culinary Professionals*. Hoboken, NJ: Wiley, 2004.

Labensky, S. R. *Applied Math for Food Service*. Upper Saddle River, NJ: Prentice-Hall, 1998.

Reinhart, P. *The Bread Baker's Apprentice: Mastering the Art of Extraordinary Bread.* Berkeley, CA: Ten Speed Press, 2001.

Carl R. Seaquist
Catherine C. Galley

Crochet and Knitting

Category: Arts, Music, and Entertainment.
Fields of Study: Geometry; Measurement; Representations.
Summary: Crochet and knitting can be used to create models of mathematical surfaces.

Crochet and knitting are techniques for turning one-dimensional yarn or thread into two-dimensional fabric by knotting it in a regular pattern. Both produce flexible, elastic fabric, although crochet is firmer than knitting. Historically, crochet and knitting were used to produce both functional and ornamental textiles by hand, but both are now hobby pursuits.

Since both techniques produce regular arrays of stitches, they can be used to display a wide variety of symmetric patterns. Furthermore, both can be used to make intrinsically curved fabrics. This allows mathematicians and others to approximate or replicate the geometry of hard-to-visualize objects, including models of two-dimensional mathematical curved surfaces, such as spheres, tori, or sections of the hyperbolic plane. Crocheting and knitting circles have been held at professional mathematics conferences for both recreation and serious discussion of mathematical concepts. Mathematician Carolyn Yackel has noted, "Knitting and crocheting are helping us think about math we already know in a different light."

Crochet

In crochet, stitches are made by pulling loops of yarn through each other with a hook. One stitch is worked at a time. Every crochet stitch is attached at its base to an earlier stitch. Varying the type of stitch and the way new stitches are worked into earlier stitches can produce many different patterns. Crochet can be worked back and forth in rows or in circular rounds. Working two stitches into one base stitch increases the number of stitches and makes the fabric wider; decreasing the number of stitches reduces the width of the fabric. Placing increases or decreases at the edges of the work makes flat fabric with curved edges. Placing increases or decreases in the middle of the fabric makes it intrinsically curved.

A hexagonal medallion made by crocheting in rounds. Crochet can be also worked in rows. (Elizabeth L. Wilmer)

The origins of crochet are not well understood. Few—if any—samples are known from before the nineteenth century. At that time, it was generally worked in fine cotton or linen thread and used for lace edgings, doilies, and other household textiles. From the middle of the twentieth century on, crochet has generally been worked in thicker yarn. It is often used to make blankets known as "afghans." The hobby of crocheting stuffed animals, known as "amigurumi," has spread around the world in recent years; because of the curved shape that these toys are crocheted in, they have few seams.

Several mathematicians have designed crocheted models of mathematical curved surfaces. As mathematician Daina Taimina has pointed out, it is especially simple to crochet negatively curved surfaces, such as a hyperbolic plane; the crocheter simply works an increase (an extra stitch) once every two or three (or n) stitches in every row. These increases cause the fabric to fold back on itself rather than lie flat. The closer together the increases are, the more ruffled the fabric.

The Hyperbolic Crochet Coral Reef, a project by the Institute for Figuring in Los Angeles, is intended to increase awareness of global warming issues by bringing together mathematicians, marine biologists, and community crafters in a highly visible way. The project asks volunteers to crochet models of coral reef life forms using Taimina's patterns. This effort and other mathematical crochet or knitting projects have been used successfully by mathematics educators in their classrooms.

Knitting

In knitting, as in crochet, stitches are made by pulling loops through each other. Knitting can also be worked in either rows or rounds. Two (or more) needles are used and many stitches are held on the needles simultaneously. The most basic stitches are "knit" and "purl" and there are techniques for increasing, decreasing, and making textural elements such as holes, cables, or bobbles. Knitting produces a flatter, stretchier fabric than crochet. (Indeed, most elastic fabric produced today is machine knitted.) As with crochet, increases and decreases allow the knitter to change the shape and curvature of the fabric. The shaping and elasticity make knitting ideal for garments such as socks, hats, gloves, and sweaters where both fit and comfort are important.

Hand knitting was once an important industry in Europe. Medieval guilds produced stunning garments for the wealthy in the Middle Ages, and a large cottage industry knitted stockings in the eighteenth and nineteenth centuries. Written patterns become available in the nineteenth century, and ornate knitting in fine thread became a popular pastime for ladies.

Hand knitting resurged in popularity in the first decade of the twenty-first century. Many current designers of garments and home textiles take their inspiration from mathematics, using symmetry and geometry to create attractive garments and household items.

Like crochet, knitting can be used to produce curved mathematical surfaces. Wide, soft, knitted Mobius bands are often knitted for use as scarves.

Further Reading

Belcastro, Sarah-Marie, and Carolyn Yackel, eds. *Making Mathematics With Needlework.* Wellesley, MA: A K Peters, 2008.

Knit and purl stitches combined to create a basketweave pattern. (Elizabeth L. Wilmer)

Bordhi, Cat. *A Treasury of Magical Knitting.* Friday Harbor, WA: Passing Paws, 2004.

Gaughan, Norah. *Knitting Nature: 39 Designs Inspired by Patterns in Nature.* New York: STC Craft/Melanie Falick Books, 2006.

Obaachan, Annie. *Amigurumi Animals: 15 Patterns and Dozens of Techniques for Creating Cute Crochet Creatures.* New York: St. Martin's Press, 2008.

Osinga, Hinke, and Bernd Krauskopf. "Crocheting the Lorenz Manifold." *Mathematical Intelligencer* 26, no. 4 (2004).

Taimina, Daina. *Crocheting Adventures With Hyperbolic Planes.* Wellesley, MA: A K Peters. 2009.

Elizabeth L. Wilmer

Dice Games

Category: Games, Sport, and Recreation.
Fields of Study: Algebra; Data Analysis and Probability; Number and Operations.
Summary: Probability is the key factor for winning any dice game.

Dice games use one or more dice as central components of the activity, which excludes board games using dice solely as random devices to determine moves. The definition can be murky, as in the case of Backgammon, dice outcomes determine a player's moves and are integral parts of game strategies. Historically, dice games involving gambling led to the creation of probability.

History

Archaeological evidence from as early as 6000 B.C.E. shows that dice games were part of early cultures, where dice were cast to invoke personal divinations. The notion of "luck" was not involved, with the dice rolls controlled by the gods. Gamblers still refer to Fortuna, the Roman goddess and Jupiter's daughter, as their "Lady Luck."

The ancient die differed from the six-sided cube bearing pips, as the number of sides varied with the materials used, including fruit pits, nut shells, pebbles, and animal knucklebones. The latter, with four sides involving different probabilities, led to the phrase "rolling the bones."

Compulsive gambling and dice games have always been connected, being traced to Egyptian pharaohs, Chinese leaders, Roman emperors, Greek elite, European academics, and English kings. On the request of professional gamblers in the fifteenth and sixteenth centuries, mathematicians such as Fra Luca Bartolomeo de Pacioli and Girolamo Cardano began to study the probabilities of winning dice games. In the seventeenth century, correspondence between Blaise Pascal and Pierre Fermat ultimately solved the "problem of points" and established basic principles of probability.

The problem of points involves a dice game between two players; multiple rounds are played with each player having an equal chance of winning on each roll. If the game was interrupted before either player had won the necessary number of rounds, gamblers could not determine the "fair" division of stakes based on current scores. Fermat and Pascal's solution analyzed the probability of dice rolls and each player winning the pot.

The dice game craps is thought to have developed from a simplification of the Old English game Hazard. (iStockphoto)

Types of Dice Games

The simplest dice game involves a single die, where the winner is the person rolling the highest number. This can be extended to rolls of multiple dice, with the player's score being the sum or product of the numbers shown. Since these dice games involve only luck, gamblers prefer variations with elements of strategy.

The dice game craps involves strategy, as the "shooter" controls the number of dice rolls and betting options. Though craps is complex, key elements can be explained. Mathematically, each roll of two dice has 36 possible outcomes with shown totals ranging from "2" to "12". However, the probabilities of the totals vary, as the probability of a "2" (known as "snake eyes") or "12" (known as "boxcars") is 1/36, while the probability of a "7" is 6/36. Prior to the first "come out roll," players bet on the "Pass Line" or "Don't Pass Line." If the "shooter" then rolls a "7" or "11," the "Pass Line" bet wins double their amount and the "Don't Pass Line" bet is lost. However, if the initial roll is a "2," "3," or "12," the "Pass Line" bet is lost, while the "Don't Pass Line" bet is doubled if a "2" or "3" shows and is returned if a "12" ("push") shows. A sum of "4," "5," "6," "8," "9," or "10" becomes the "point" number, which the shooter tries to duplicate on the second roll. If the point number is made, the point bet is won and additional rolls can be made. But, if a "7" is rolled before the point number, the shooter "craps out" and a different shooter starts a new round. Craps games involve many other options, such as "Come/Don't Come Bets" and "Horn Bets."

Other dice games are used for gambling, each with their own multiple versions and strategies. For example, in the dice game Ship, Captain, and Crew, a players gets three rolls of five dice to gain a ship ("6"), a captain ("5"), and a crew ("4") in that order (or simultaneously). When those special numbers are rolled, that die is removed from play, with a successful player's score being the sum of a roll of the two remaining dice.

In Buck Dice, a player throws one die to determine the "point number." Another player then rolls three dice, continuing the rolls as long as one of the dice equals the point number. When this doesn't occur, the player's score for that round is the number of rolls. A "big buck" occurs when all three dice equal the point number, and the player withdraws from the game. A "little buck" occurs if all three die do not equal the point number, which adds 5 points to the player's score. Any player with exactly 15 points withdraws from the game; any score forced higher than 15 nullifies a roll, and the player must reroll. The loser is the last person without reaching 15.

In Aces, a player starts with at least five dice, which he or she loses according to the numbers thrown. All rolled "1"s are placed in the table's center and eliminated. All rolled "2"s are passed to the player on the left, while all "5"s are passed to the player on the right. Turns continue with rolls of the remaining dice until players either do not throw a "1," "2," or "5," or have lost all of their dice. Play continues around the table until the last die rolled is a "1," and the player who threw it is the winner.

Farkle begins with a player rolling six dice. Each "1" adds 100 points, each "5" adds 50 points, and if three dice show the same number, the player adds 100 times that shown number. A player can stop after any roll and keep the current total. Alternately, a player can roll again to possibly increase his or her score. But, if the next dice do not produce a positive score, the player lose all accumulated points for that round. The winner is the first to reach 10,000 points. Some variations of Farkle give 1000 points for shown runs of "1–5" or "2–6."

In line with their history, multiple versions of dice games exist and will continue to be used by gamblers. Thus, the players who understand the probabilities involved will always have the advantage.

Further Reading
Barboianu, Catalin. *Probability Guide to Gambling: The Mathematics of Dice, Slots, Roulette, Baccarat, Blackjack, Poker, Lottery, and Sport Bets*. Craiova, Romania: INFAROM Publishing, 2008.
Bell, R. C. *Board and Table Games From Many Civilizations*. New York: Dover Publications, 1979.
Devlin, Keith. *The Unfinished Game*. New York: Basic Books, 2008.
Mohr, Merilyn. *The Game Treasury*. Shelburne, VT: Chapters Publishing, 1993.

JERRY JOHNSON

Extreme Sports

Category: Games, Sport, and Recreation.
Fields of Study: Algebra; Geometry.
Summary: The emphasis on fast motion, tricks, and personal expression in extreme sports makes geometry especially relevant to athletes.

There is no single definition of extreme sports, though they generally include dangerous sporting activities that involve a substantial risk of injury, like Buildings, Antennae, Spans, and Earth (BASE) jumping, cliff diving, street luge, or even the traditional running of the bulls in Pamplona, Spain. Extreme sports are believed to be attractive to participants because of the challenge and adrenaline rush and to spectators because the results are typically unpredictable.

The popularity of extreme sports grew rapidly in the latter part of the twentieth century. The television network ESPN created the Extreme Games, now called the "X Games," in 1995, making extreme sports more visible to the general public. Other networks have also begun to televise these types of competitions and some extreme sports events have been included in the Olympic Games. Mathematics is important in extreme sports. Knowing and applying concepts from geometry and probability helps participants be safe and successful. Innovative equipment manufacturers use concepts and techniques from many areas, including geometry,

When a skateboarder performs an ollie, the forces acting on the board are the weight of the rider, the force of gravity on the board, and the force of the ground pushing up on the board, which balance out to zero net force. (iStockphoto)

statistics, modeling, and simulation, to prototype and refine their designs, resulting in greater safety and effectiveness.

Skateboarding

Skateboarders perform tricks using a wheeled board, either on a flat surface or using equipment like ramps or rails. Many stunts rely on differential pressure applied by the rider's feet to various parts of the skateboard to tilt or flip it, often rotating both board and rider in one or more axes. Lip tricks require a vertical orientation and transitional edge like the lip of a swimming pool or ramp. In aerial tricks, the rider leaves the ground completely, using counterpressure of hands and feet to maintain control of the board while spinning or flipping.

Tony Hawk is one of the most well-known extreme athletes and a vertical skateboarding pioneer. He was the first person to competitively perform an aerial turn of two and a half rotations, or 900 degrees, at the 1999 X Games. In the past, he has done 720 degree turns. For the 900, he exerted greater takeoff force in the direction of the turn, producing more rotational velocity. Tony Hawk's Project 8 video game used motion capture technology to smoothly animate professional skaters, while Tony Hawk Ride allowed players to simulate the sport using a skateboard-like controller.

Snowboarding

Snowboarding is similar to skateboarding and involves standing on a board and sliding down a snow-covered hill. Snowboarding became an Olympic sport in 1998, with giant slalom and half pipe competitions taking place. The giant slalom is a speed race in which athletes speed down a steep hill with gates that require them to zigzag between. Determining an optimal path from one gate to another without crashing or wasting time requires mathematics, especially geometry. A half-pipe consists of two quarter-cylinders connected by a flat space and topped by a small lip. The competition is a more artistic event, with athletes generating enough speed using the curves of the pipe to become airborne and do tricks. These may include multiple rotations, both twisting and somersaulting. At the 2010 Olympics, Shaun White executed a record-setting 1260-degree trick consisting of two flips and three and a half spins.

BMX Biking

In bicycle motocross (BMX), athletes ride specially designed smaller bicycles that enable them to shift their center of mass to make precision movements. BMX courses often use steep hills to launch the rider into the air to perform tricks. Other tricks and spins may be done on flat ground. The sport was added to the list of events for the 2012 Summer Olympic Games. Billy Gawrych is a professional BMX competitor who performs intricate routines, often set to music, with tricks linked together in a series of connected, flowing patterns.

Sports Engineering and Equipment

Sports engineering is a growing interdisciplinary field that draws from mathematics, engineering, biology, physics, materials science, and many other disciplines to study characteristics of athletes and equipment, as well as their interaction. The focus is on performance and safety. For example, engineer Mont Hubbard described the motion of skateboards with riders using two mathematical models, and mathematicians develop new models using techniques and theories from areas like trigonometry, physics, differential equations, and probability. Quality function deployment is a method of quality control that attempts to translate often subjective customer requirements into mathematical engineering specifications. One research group studied the subjective perception of the "feel" of snowboards. They used field evaluations and laboratory data to create matrices of parameters. Snowboards for freeride and freestyle, the two primary types of snowboarding, have somewhat different designs; however, issues of flexibility, torsional stiffness, and curvature were the important factors affecting feel and performance for both styles. Equipment for sports of all kinds is subjected to statistically designed tests to evaluate safety, and data from accidents and failures helps fuel further research.

Further Reading

Clemson, Wendy, David Clemson, Oli Cundale, Laura Berry, and Matt King. *Using Math to Conquer Extreme Sports*. New York: Gareth Stevens Publishing, 2004.

Estivalet, Margaret, and Pierre Brisson. *The Engineering of Sport 7*. Vol. 1 New York: Springer, 2008.

Gutman, Bill. *Being Extreme: Thrills and Dangers in the World of High-Risk Sports*. New York: Citadel Press, 2003.

Sagert, Kelly Boyer. *Encyclopedia of Extreme Sports*. Westport, CT: Greenwood Press, 2008.

Thorpe, Holly. *Snowboarding Bodies in Theory and Practice (Global Culture and Sport)*. New York: Palgrave Macmillan, 2011.

Tyler, M., and K. Tyler. *Extreme Math: Real Math, Real People, Real Sports*. Waco, TX: Prufrock Press, 2003.

Michele LeBlanc
Nena Amundson

Fantasy Sports Leagues

Category: Games, Sport, and Recreation.
Fields of Study: Algebra; Data Analysis and Probability.
Summary: Fantasy sports leagues employ a variety of algorithms to predict player performance and to rank players and teams within each league.

In fantasy sports leagues, players act as the owners and managers of virtual sports teams that are typically composed of real players who are active in a given sport during a competitive season. Performance statistics for individual athletes on a fantasy owner's roster, who usually belong to many different teams in the sport in real life, are mathematically combined to produce "fantasy points" for the owner. Often, owners may trade athletes or must make other types of decisions about who on their roster will be counted as "active" for a given period of competition, just like real managers. Fantasy baseball and fantasy football have historically been the most popular but fantasy leagues have evolved for many other sports, including basketball, golf, hockey, soccer, auto racing, and even cricket. Different leagues, even within the same sport, use a variety of formats, statistics, and weighting schemes to compute fantasy points. Season winners are usually the owners who have accrued the most fantasy points. While such games have existed in one form or another since at least the end of World War II, the development of the Internet drastically changed the nature and popularity of fantasy sports leagues by providing real-time access to data and tools for automated computation, making the activity more accessible for a broader range of participants. There are estimated to be millions of fantasy sports players in the United States alone. In the twenty-first century, mathematicians and others study fantasy sports leagues, and they have become a tool in mathematics classrooms as well.

History

Fantasy sports leagues grew from other types of sports simulator games that used data from past seasons and random number generation to determine the outcomes of simulated games. One of these was Strat-O-Matic, a board game using player statistics cards and dice that was developed by Hal Richman. It premiered in 1961 and still exists in both card and computerized forms. Richman began developing the game as a child because he "loved baseball and numbers" and disliked what he saw as unrealistic randomness in other baseball board games. He released the game while earning his undergraduate mathematics degree. John Burgeson, an IBM computer programmer, created a computer fantasy baseball simulator in 1960 that used random numbers and player statistics to generate a play-by-play description of a game between two teams. Many real baseball managers reportedly played fantasy-style games when they were young. According to writer Alan Schwartz, "That's how they learned how to apply the mathematics of risk-taking."

In the 1970s and 1980s, early fantasy leagues began to emerge for baseball and football. Writer Daniel Okrent developed Rotisserie League Baseball, which was named after the restaurant where players conducted the first draft. "Rotisserie baseball" is now a standard term for this widely-used format. It differed from most older games by using current-season statistics and data as they occurred rather than past seasons' statistics. This style of play became popular after an *Inside Sports* magazine article described the rules of the game and discussed the league's first season. Statistician George William "Bill" James also developed the analytical methodology of sabermetrics around this time and his *Bill James Baseball Abstract* was widely used by fantasy players. Similar mathematical analyses were produced for fantasy football by *Fantasy Football* magazine, which evolved into the print and online *Fantasy Football Index* (and also *Fantasy Baseball Index*). These publications and many others provided mathematically modeled variables, such as dollar values, statistical projections, and optimization strategies, for fantasy players. Sometimes the modeling proved useful enough that the writers went on to advise real teams.

Before the Internet, coordinating fantasy sports and calculating points could be time consuming. Data came largely from print sources, which were time delayed. A standard 162-game baseball schedule required near-daily computations for each owner in the league. Fantasy football was somewhat less challenging because of the smaller number of games in a season, but most fantasy methods had to restrict the number of variables used. Some commercial statistical services started to fill this need by compiling databases of sports statistics and providing services to calculate points—for a fee. Results were mailed or faxed; later, they could be sent electronically. The development of the World Wide Web in the mid-1990s facilitated and often automated the process of tracking player statistics and calculating points and league standings. Fantasy players could also quickly communicate with each other using e-mail, message boards, and chat rooms, resulting in online communities and worldwide leagues. Researchers have modeled this growth using sociologist Everett Rogers's diffusion of innovation theory. The curve of fantasy players over time exhibits the classic *S*-shape of slow initial growth among early innovators and adopters, a middle period of accelerated growth, and a saturation of the market leading to a leveling off or slower growth period. The rapid growth of fantasy sports in the late twentieth century led to issues related to its potential classification as gambling, fairness in prizes, and the legal rights of players or teams to control the dissemination and use of statistical information about professional athletes, especially when outside companies were making a profit from such use.

Mathematical and Social Connections

The line between fantasy sports and real sports is often blurred and mathematical methods used in one are often applied to the other. For example, mathematicians have explored a concept often called the "magic number" or elimination number, which quantifies the number of games a team must win to avoid being eliminated from the championship. The problem is popular in computer science classes. A common solution is to compare the number of games a team has left to play to the win-loss difference of the nearest rival. Researchers found that the numbers for all teams may be found simultaneously as they are a function of the number of games won plus the number of games left to play. Other mathematicians investigate optimal strategies for drafting players to teams using methods such as stochastic dynamic programming and deterministic dynamic programming coupled with various types of mathematical modeling and decision making. Some have researched the extent to which players rely on mathematical modeling and statistical methods instead of on heuristics and personal preferences. Mathematics teachers have found some success in using fantasy sports to motivate students and to help them succeed. Additional evidence suggests that fantasy sports may help reduce gender gaps in mathematics achievement. Some girls have stated that fantasy sports are "cool" and help them relate to boys as equals, and women are involved in the creation and management of fantasy leagues. For example, Jordan Zucker, who has an undergraduate degree in mathematics, created the *Girls' Guide to Fantasy Football* Web site and manages an all-female fantasy football league.

Further Reading

Fantasy Sports and Mathematics. http://www.fantasy sportsmath.com.

Fry, Michael, Andrew Lundberg, and Jeffrey Ohlmann. "A Player Selection Heuristic for a Sports League Draft." *Journal of Quantitative Analysis in Sports* 3, no. 2 (2007).

James, Bill. *The Bill James Handbook 2011*. Chicago: ACTA Publications, 2010.

Schwarz, Alan. *The Numbers Game: Baseball's Lifelong Fascination with Statistics*. New York: St. Martin's Press, 2004.

Sarah J. Greenwald
Jill E. Thomley

Fireworks

Category: Architecture and Engineering.
Fields of Study: Algebra; Geometry; Number and Operations.
Summary: Firework mathematics involves the timing and rhythm of burning, rocket flight, and explosions.

Fireworks are explosions for entertainment with design elements of light, sound, and smoke. Chemical additives are used to color fireworks, which originated in ancient China. The province of Liuyang is known as the home of fireworks. Fireworks as an art are temporal, like dance or animation; therefore, much of firework mathematics has to do with the timing and rhythm of burning, rocket flight, and explosions. Mathematicians around the world have modeled and quantified various aspects related to fireworks, like the path and maximum height. In the seventeenth century, Claude Dechales published what became a popular textbook on mathematics that included pyrotechnics. Engineer Amédée-François Frézier, whom some also refer to as a mathematician, worked on the theory of fireworks in the eighteenth century. The process of mathematical induction has been likened to a sequence of connected fireworks. In the United States, the Bureau of Alcohol, Tobacco, Firearms and Explosives classifies and regulates fireworks.

Patterns of Explosions

Most fireworks shot into the air explode in spherical patterns. By modifying the composition of fireworks, it is possible to add or remove tail effects, change the speed of individual parts, and produce delayed explosions to parts, filling spheres with radial lines or creating expanding spheres. Less frequent are fireworks that burn sustained, extending, two-dimensional shapes, such as rings or hearts.

Ratios and Proportions of Shells and Mortars

Many fireworks are packed into shells and fired out of special mortars, or small cannons. Larger shells are fired out of larger mortars with higher speeds and also fly higher. As with any projectile, the path and height of the firework shell, until the explosion, obey the quadratic equation of gravitational deceleration and the shell flies following the path of a parabola. On the other hand, because of the physics of the black powder or pyrex used to propel the shells out of the mortars, the relationship between the size of shells, mortars, and their initial speed is linear. The relationship between the size of the shells and the maximum height they fly is also linear. Pyrotechnician formulas approximate 100 feet of the shell's maximum flight height per every inch of its diameter. The explosion of the firework has to be timed so it happens when the shell is high up in the air, which is achieved through solving the height equation and matching the time of chemical reactions in the shell to that height.

Fireworks Color and Temperature Gradients

There are two distinct ways to color fireworks. The first method is based on the same physical process used in incandescent light bulbs and the second on that used in neon lights. The first method uses blackbody radiation—the property of objects to emit more light with higher temperature. Blackbody radiation emits light over a broad spectrum. As metals heat, they start to become red to the human eye because the majority of the spectrum is light at infrared wavelengths human beings cannot see. When the temperature rises, the emission of the light in the visible spectrum increases and the object becomes first yellowish and then white, the mixture of all visible-light wavelengths. Thus, fireworks that depend on blackbody radiation for their color can only be dull red, pale yellow, or white.

The second method of firework coloring is based on the so-called atomic emission. Atoms in the firework material, before the firework is fired, are in a stable state, corresponding to particular orbits of electrons. If atoms are electronically excited, they emit photons to return to that stable state. When photons are in the visible spectrum, the human eye sees a color as the atomic emission takes place. Some elements have a narrow spectral band in their atomic emissions, allowing particular pure colors to be pinpointed. For example, sodium emits bright yellow and barium emits green when electronically excited. Copper salts emit pure blue but they are so unstable at high temperature that people only recently learned to use them safely in fireworks.

If the firework material burns too hot, the blackbody radiation process takes over. Therefore, to produce pure colors of the atomic emission process, pyrotechnicians create mixtures that burn relatively cool. The chemistry breakthrough allowing this to happen was

Many of today's fireworks are made in much the same way they were hundreds of years ago. (iStockphoto)

the substitution of potassium chlorate, which burns at around 120 degrees Celsius, for potassium nitrate, which burns at 560 degrees Celsius. Fireworks contain coolants that prevent burning from reaching higher temperatures, for example, by releasing some water and carbon dioxide, as sodium bicarbonate does.

Pyrotechnic Competition and Measurements

At competitive events, fireworks are measured based on several criteria, mostly qualitative and artistic. The quantitative criteria include purity and brightness of color and the appropriate explosion height. The timing of the intended fireworks effects, such as the change of shape and color, is also taken into consideration—it has to follow a recognizable temporal pattern and to form a pleasing rhythm.

Competition judges add points for technical difficulty, celebrating innovations in fireworks. For example, when strobe effects were first discovered, fireworks using them were awarded technical difficulty points at competitions. After a few years, as strobe effects became well researched, judges stopped awarding points for them.

Further Reading

Danby, J. M. A. "Fireworks." *The College Mathematics Journal* 23, no. 3 (1992).

Lancaster, Ronald. *Fireworks, Principles and Practice*. 4th ed. Gloucester, MA: Chemical Publishing, 2005.

Shimizu, Takeo. *Fireworks: The Art, Science, and Technique*. 3rd ed. Post Falls, ID: Pyrotechnics Publications, 1996.

Maria Droujkova

Fishing

Category: Games, Sport, and Recreation.
Fields of Study: Algebra; Geometry.
Summary: Fishing tactics, management, and measuring all require the sophisticated use of mathematical principles.

Mathematics has proven to be a useful tool in understanding the impact of a variety of factors that influence fish populations. Other mathematical techniques have been used to analyze photographs of fish and to generate useful estimates of the fish's weight. Mathematics has also demonstrated its utility in the creation of tools for locating and catching fish.

Fishery Management

The estimation and regulation of the striped bass and bluefin tuna populations along the East Coast of the United States are examples of important fishery management issues with serious economic implications.

Mathematics as an ecosystem-based management tool has been used to formulate population models that attempt to account for very complex environmental factors, including variations in water quality and temperature; fluctuations in the availability of important forage species upon which the targeted species depend for food; the presence (or lack thereof) of appropriate spawning areas; the impact of fish farming on wild fish populations; the interplay of commercial fishing and sport fishing; the introduction of invasive species; and the impact of diseases. Regulations regarding the timing, size, and number of fish that are to be harvested are based, in part, on mathematical models. Presumably, understanding the likely consequences of changes in these and other factors will lead to improved management decisions. An alternative management approach has been suggested by analysis of the history of the sardine fishery in California's coastal waters. Such evidence has led some mathematicians to believe that fluctuations in fish populations are best explained by utilizing branches of mathematics known as "complexity theory" and "chaos theory."

Weight Estimation

Mathematicians were called upon when the National Freshwater Fishing Hall of Fame faced controversy over its listing of the record muskellunge as a fish caught in 1949, reported to be 63.5 inches in length and weighing 69 pounds. Three photographs of the angler holding the fish in front of him documented the catch. The question arose, since the height of the angler in the photograph was known: could the length of the fish be accurately estimated? In fact, projective geometry together with some precise measurements gleaned from the photographs could provide very good estimates of the length of the fish. However, a difficulty remained: was there a way of accurately estimating the weight of a muskellunge based upon its length, without knowing its girth? In fact, an algebraic formula has been developed for estimating the weight of a muskellunge that requires only a precise measurement of the length of a portion of the fish's body. The formula is

$$W = \frac{L^3}{2800}$$

where W is the weight in pounds and L is the length in inches.

Tools for Locating and Catching Fish
The electronic devices often utilized in locating fish include flashers, LCD graphs, and global positioning systems. Each of these items depends upon mathematical underpinnings. However, mathematics also plays an important role in the creation of the nonelectronic tools used in sport fishing.

The design of reels, fly lines, and fishing rods depends upon mathematics. The role of geometry is especially apparent in the building of traditional split-bamboo fly rods. For example, in a two-piece split-bamboo rod, each of the two sections of the rod requires that six strips of bamboo be cut and planed to a precise taper such that each strip has cross sections along its length that are equilateral triangles of diminishing size. When these strips are properly glued together, hexagonal cross sections result. The rod blank so created is the foundation of a bamboo fly rod. The builder must still decide where to place the line guides along the length of the blank in order to produce a fishing rod that will both cast well and enable the fisherman to quickly capture hooked fish. Not only does the distance between consecutive guides increase from the rod tip toward the butt of the rod but also those distances change in a precise way. The initial placement of the guides on the rod is accomplished by using an idea from algebra known as "arithmetic progression." The fine-tuning of the guide placement on the rod then depends upon measuring the arc through which the rod bends when placed under a predetermined load.

Further Reading
Raeburn, Paul. "Using Chaos Theory to Revitalize Fisheries." *Scientific American* (February 2009).
Yami, Ben. "Mathematics and Selective Fishing." *WorldFishing & Aquaculture* (June 1, 2009).

Philip McCartney

Football

Category: Games, Sport, and Recreation.
Fields of Study: Data Analysis and Probability; Geometry; Measurement.
Summary: Football coaches use statistics to inform their decisions while the National Football League analyzes the effects of its rules.

Though a physical battle between two teams of talented athletes, football can be analyzed using mathematical ideas and techniques. Pertinent to coaches, players, fans, and betting agents, these analyses focus on all aspects of football—the physical aspects and performance of players, game elements (passing, running, defense, and kicking, as well as strategies), and the geometry-based physics surrounding the game. Mathematical analysis can impact the game positively or negatively. Nonetheless, football remains a physical competition between two teams, despite the use of mathematics to identify patterns of strengths and weaknesses, suggest optimal strategies, provide rankings, stimulate discussions, and possibly resolve arguments.

Quarterback Rating
The National Football League uses a mathematical formula to rate quarterbacks. Data are collected for each game and for the season relative to a quarterback's pass completion percentage (P), touchdown pass percentage (T), pass interception percentage (I), and average gain per attempt (G). Using a few boundary conditions, a quarterback's rating (Q) is determined by the formula

$$Q = \frac{5}{6}(P + 4T - 5I + 5G + 2.5).$$

The formula's derivation in terms of four independent variables involves multiple regression techniques.

Overtime Rules
The National Football League also uses Markov chain techniques to analyze its overtime rules in response to the "statistical fact" that too many football teams were winning important games with a field goal on their first overtime possession. Thus, the "winning" team, after a hard-fought game, is influenced too greatly by a single coin flip that determines team possession, with minimal differences accounted for by a team's ability to score on the first possession. Effective in 2011, the

rules for play-off games were changed to prevent the game ending with a field goal on the first possession of overtime.

Though difficult to implement practically, geometry, trigonometry, and calculus all play strong roles within a football game and its situations. Examples include the following:

- Use of a quarterback's physical characteristics to determine the best angle and release points for throwing a pass, assuming it must reach receivers in different field locations and at multiple distances
- Use of the law of cosines to both understand and improve passing angles, timing, and patterns run by receivers
- Determination of an optimal efficiency for punters on each kick, or the ratio of the actual kick's distance to the maximum possible distance using the same force
- Prior to kicking a field goal, determination of success in terms of the angle subtended by the two goal posts
- Determinination of a defensive lineman's stance to maximize centers of gravity and potential force on impact with an opposing linemen

By gathering and analyzing the available data provided by a game, probabilities can help examine the particular events happening within a game, such as the following:

- Likelihood of a team making 0, 1, or 3 points after a touchdown score
- Reality of a quarterback having a "hot hand" in his or her completion of successive passes
- Success of making a field goal, given it will or will not result in a change in who has the leading score
- Probability of scoring during a fourth-and-goal
- Monitoring a coach's decisions in calling plays, especially if "conservative"
- Probability of a record being broken, either by a team or by a player

Similarly, mathematical statistics provide perspectives that explain game occurrences, provide comparative rankings of teams and players, and assist in decision making by coaches and team management. The usual sources of statistics are data regarding passing, running, defense, kicking, turnovers, and time management. Examples include the following:

- Use of ratios, means, and medians as descriptive statistics for a player, a position, a game, or a season
- Use of logistic regression models to calculate end-of-game point differentials, based on independent variables such as turnovers, passing yardage, running yardage, penalty yardage, number of first downs, and number of completed passes
- Impact of icing a place-kicker at crucial times within a game
- Correlations between a player's characteristics and training regimens relative to game performance
- Trend analysis, based on either a player's or a team's performance in specific ways over the past 5, 10, and 15 games
- Winning tendencies based on connections to lead changes during a game or knowledge of the team leading at the end of the third quarter
- Impact of rules changes on team scoring and defenses within the sport itself, such as observed effects of initial field positions subject to penalties or punts out of bounds
- Determining the "best" all-time player in a particular position (for example, quarterback, tight end, halfback, linebacker, or field-goal kicker), at a particular time in a game (such as the last quarter) or in an era
- The use of digraphs and "mysterious" statistical formulas to determine weekly rankings and placement of teams in a bracketed tournament (such as the Bowl Championship Series), directly affecting betting pools with stated odds
- Selection of players by professional teams during the annual draft, using historical data for each player's performance in conjunction with physical data

- The use of statistical data as part of contract negotiations between players and management, or even the release or trading of players based on team needs
- The questionable yet significant correlation between stock market performance and the Super Bowl's winning team

Mathematical game theory is evident in a coach's decision-making process, such as on each play within a football game, hoping to choose optimal tactics. The specific decisions range considerably and include the following:

- A coach's choice of designed offensive plays and defensive set-ups, relative to the down, position on the field, time of game, score, and opponent
- A coach's calling of time-outs and play reviews at opportune times
- A coach's use of techniques to motivate specific players
- A team's selection of players during a draft, dependent on the players' apparent abilities, the inferred needs of other teams, and the specific draft round
- Contract negotiations involving players, agents, and team management

Finally, using these statistical data and mathematical modeling techniques, one can create realistic simulations of football games, possibly using computer animations.

At the collegiate and professional levels, coaches increasingly use mathematics to remain competitive, even hiring mathematical statisticians as important parts of their staff. However, some authors and fans suggest that the football team with the best players and coaching will usually win, despite any use of sophisticated mathematics.

Further Reading

Bennett, Jay, and James Cochran. *Anthology of Statistics in Sports*. Philadelphia, PA: Society for Industrial and Applied Mathematics, 2005.

De Mestre, Neville. *The Mathematics of Projectiles in Sport*. New York: Cambridge University Press, 1990.

Eastway, Rob, and John Haigh. *Beating the Odds: The Hidden Mathematics of Sport*. London: Robson Books, 2007.

Friedman, Arthur. *The World of Sports Statistics: How the Fans and Professionals Record, Compile, and Use Information*. New York: Athenaeum, 1978.

Gay, Timothy. *The Physics of Football*. New York: HarperCollins, 2005.

Jerry Johnson

Geometry in Society

Category: School and Society.
Fields of Study: Connections; Geometry.
Summary: Geometry permeates society from its many applications in daily life to its usefulness as a framework for deductive inquiry.

Geometry has long been useful in society for both practical purposes and as deductive inquiry. The word itself is a combination of two ancient Greek words: *geo* (Earth) and *metron* (a measure).

Thus, a direct translation might be "Earth-measuring." Geometry developed from practical needs in ancient cultures, such as the taxation of lands and the construction of monuments. In many settings, geometry played an important role in both aesthetic quality and stability. For instance, in art and architecture, beautiful geometric figures tiled surfaces. Stability notions, like the center of mass, could be calculated using geometry, and a camera's tripod has three legs because three points determined a plane, so three legs made it more convenient to find a stable position on an arbitrary surface. The Greeks explored geometry as an axiomatic system, and for thousands of years geometry was an essential part of a liberal arts education. Along with fields such as algebra and analysis, it also formed a core area in research. However, the role of geometry in school has changed over time, reflecting the priorities of society, researchers, and industry. In addition, educators have long debated which geometric topics should be taught. In some college settings in the twentieth century, the prominence of geometry declined. Some topics from courses like discrete geometry were taught in other departments, like computer science.

Emerging fields, like algebraic geometry, were associated with algebra programs. While geometry was no longer a core area in some undergraduate mathematics curricula, it remained important in all levels of school in one way or another because it could be used in so many occupations. Construction, design, and architecture are just a few of the jobs that make use of geometry.

Early History

One development from the history of geometry and measurements of length, area, and volume can be found about 3000 years ago, when peoples in ancient Egypt farmed along the Nile River. King Sesostris is noted as having divided the land into rectangles. He taxed farmers based on the area of the land they occupied. But there was a problem: every year, the Nile River flooded the surrounding area. After flooding, a large portion of the lands allocated to farmers was destroyed. Hence, Sesostris had to exempt the tax on the destroyed lands. To do this, he had to measure the exact area of destroyed land. Another problem that naturally arose was how to divide the land among a number of farmers. Covering a given region by pieces is called a "tessellation" or a "tiling" of the region. Precisely, a tessellation of the plane is a set of plane figures or tiles that cover the plane without any overlaps and gaps. Tessellations were also found in mosaics as well as in floor and wall coverings.

Historians theorize that axiomatic investigations arose in ancient Greece because there was a prevalence of debate and justification in Greek society. However, even though the Greeks are noted as transforming geometry into a deductive branch of mathematics, they were still interested in practical applications. Plato is noted as believing that "for the better apprehension of any branch of knowledge, it makes all the difference whether a man has a grasp of geometry or not."

Geometry Education Since the Seventeenth Century

Ideas about the utility of geometry have spurred some changes in the way that geometry has been taught over the years. Most students who went to college prior to 1800 came from some type of preparatory school or had private tutors. As more universities opened in the United States and Europe, the preparation of the students needed to be considered. In some locations, Euclidean geometry was taught directly from a translation of Euclid of Alexandria's *Elements* as a second-year course in college. Students were expected to learn how to prove everything in the *Elements* in the same way that Euclid had outlined the proof. This process developed a strong sense of proof and logical structure in the student, but may not have prepared students for geometric problems that fell out of the direct line of proofs in the *Elements*. The argument about the utility versus the deductive nature of geometry was reflected in the diverse foci of geometry education around the world: was geometry to prepare students in the formal axiomatic method offered by geometry, or was geometry to teach students about how geometry could be used? In some locations, solid geometry and spherical geometry and trigonometry for surveying and navigation were the focus, while in others, it was Euclid's planar geometry and axiomatic perspectives.

In 1794, Adrien-Marie Legendre wrote a textbook in which he rearranged the material from Euclid and added other concepts, such as measurement. This textbook was adopted by Claude Crozet and brought to the United States Military Academy (USMA) in 1817. In 1819 Charles Davies, a mathematics professor at the USMA, translated this textbook into English and started making changes to include the type of geometry useful in mensuration and navigation that the United States Army and Navy wanted of its leaders. This type of geometry was adopted by most of the other military schools in the United States. This course came to be known as "descriptive geometry," which then led to engineering drawing. By the 1840s, universities decided that students who desired entrance needed to have had a course in Euclidean geometry in high school. This requirement moved the course in geometry into the K–12 curriculum.

Bernhard Riemann, for whom "Riemannian geometry" is named, considered that: "It is well known that geometry presupposes not only the concept of space but also the first fundamental notions for constructions in space as given in advance. It only gives nominal definitions for them, while the essential means of determining them appear in the form of axioms. The relationship of these presumptions is left in the dark; one sees neither whether and in how far their connection is necessary, nor a priori whether it is possible. From Euclid to Legendre, to name the most renowned of modern writers on geometry, this darkness has been lifted neither by the mathematicians nor the philosophers who have laboured upon it." At the turn of the twentieth century, Felix Klein, who revolutionized the understanding of geometric spaces by investigating them through their transformations or symmetries, noted: "Everyone who understands the subject will agree that even the basis on which the scientific explanation of nature rests is intelligible only to those who have learned at least the elements of the differential and integral calculus, as well as analytical geometry." The debate about how geometry should be taught has continued into the twenty-first century.

Applications

Geometry is a broad subject, hence it casts a broad shadow. Henri Poincaré, whose name is attached to the Poincaré disk in hyperbolic geometry, stated "by natural selection our mind has adapted itself to the conditions of the external world. It has adopted the geometry most advantageous to the species or, in other words, the most convenient. Geometry is not true, it is advantageous." Recent work has shown that geometry may be innate and form some core knowledge in the brain. For example, some researchers have reported that indigenous tribes in the Amazon River basin have a much deeper geometric intuition—without any formal education—than Western schoolchildren. Studies of animals, including fish and chimpanzees, have indicated that they may have a Euclidean map of their home territory in their brains. There are several types of geometry that illustrate the wide variety of applications:

- *Euclidean plane geometry* is the plane geometry of Euclid. It has close connections with computational geometry, computer graphics, discrete geometry, and some areas of combinatorics. It is the geometry of engineering drawing and architecture.
- *Euclidean solid geometry* describes three-dimensional space. It is used in solid modeling, constructive solid geometry, computer graphics, engineering design, and architectural design, among other fields.
- *Differential geometry* has become increasingly important to mathematical physics and cosmology because of the work of Albert Einstein on general relativity. The objects that are considered in differential geometry are smooth objects—objects without sharp corners or edges. Differential geometry is used in econometrics in economy; to solve problems in digital signal processing in engineering; to analyze and describe geologic structures in geology; to analyze shapes in computer vision; and to analyze and process data in image processing.
- *Discrete geometry* focuses on the properties of finite or discrete objects, like lattice points. It is used in robotics, computer graphics, crystalline theory, packing theory, and configurations of objects, among others.
- *Computational geometry* is a field that includes researchers from computer science and mathematics and investigates algorithms, data structures, and computational issues related to geometric structures and operations. It is used in robotics, computer graphics, geographic information systems (GIS), computer-aided design, medicine, and machine learning, among others.

Geometry in Design and Manufacturing

Geometry is used in the planning, layout, and production of most items that are manufactured. The design process may involve finding the optimal way to lay out a pattern on a piece of cloth or on a piece of wood, plastic, or metal so as to minimize waste. Computers are used to find the best place to divide large sheets of wood in the manufacture of cabinets, flooring, and paneling so as to generate the maximal use from that wood. Areas that use geometry in this manner are quite diverse and include the following:

Architecture includes home planning, interior design, and landscape architecture.

Assembly planning involves objects manufactured using an automated assembly line or robotic manipulations. In robotic manufacturing, the constraints of the robots determine the motions that can be made, and the motions can determine the programming and design of the robotics to be used.

Computer-aided design (CAD) includes many commercial and open source programs used by architectural and manufacturing firms to complete the design of items from motherboards to cars.

Grasping and fixturing answers the question: where does one place obstacles, such as robot fingers or fixtures, to prevent some object from moving?

Machinists are professionals who work on computer numerical control (CNC) machines to make parts in the manufacturing process. They can understand the process better if they have a deeper understanding of solid geometry. The cutter on one of these machines is controlled by the computer that is reading from a design that has been programmed—probably digitized and programmed. Because of the manner in which the machine operates, most instructions do not come from reading in the standard Cartesian coordinate system, but in cylindrical or spherical coordinates, or at times in a newly developed coordinate system designed just for that machine. The tool and die makers for manufacturers across the nation must take designs—and sometimes the designs are only outlines—from the engineer and create a prototype for the part. These prototypes can now be designed in the computer using CAD and then printed on a three-dimensional printer. The geometry for "printing" these parts is complicated, but allows for faster prototyping and manufacture.

Geometry in Graphics and Visualization

Computer graphics is an area that continues to expand from its beginnings attempting to represent geometric shapes (consider the 1982 movie *Tron*) to the extensive work of Pixar and other computer-generated imagery (CGI) groups in the movie industry to bring to life entire worlds that look realistic (consider the 2009 movie *Avatar*). Shapes and figures are first designed, digitized, and then rendered as nets. Once the basic figure is digitized, it is manipulated by computers according to the movie script. Once the entire script is done, the figures are finalized to give them a more realistic appeal. Advances in this area seem relatively simple, yet the example of making Sulley's hair move realistically in the 2001 movie *Monsters, Inc.* or the realistic appearance of the water in the 2005 movie *Madagascar* took a great deal of effort to develop.

Printing and the graphic arts involve issues of layout and form. The optimal use of geometric shapes on a page or palette, relative size of objects, and perspective are some of the relevant geometric considerations.

Geometry in Information Systems

Cartography and geographic information systems (GIS) are used by most local and state governments in the United States for maintaining property and road records and for making maps.

Voronoi diagrams answers the question: given a collection of objects (for example, fire stations) to be located throughout a city, how does one allocate these objects so that each person in the city is closer to one than any of the others? The Voronoi diagram, named for Georgy Voronoi, is a geometric partition of a space. Voronoi diagrams are used in situations such as models of crystal and cell growth, locations of limited facilities, and reservoir simulations.

Geometry in Medicine and Biology

Protein and virus modeling investigates the shape of a protein or virus and its motions, which are important in understanding its behavior and in developing treatments.

Medical imaging uses lower dimensional information, such as two-dimensional images, to reconstruct the shapes of organs, bones, or tumors. The reconstruction of three-dimensional shapes from slices is a geometric problem.

Geometry in Physical Sciences

Astronomy is one of the oldest uses for solid geometry. Computational geometry problems come about in observation planning and shape reconstruction of irregular shapes, such as asteroids.

Scientific computation involves the application of computer visualization and simulation.

Physics has long been intertwined with geometry. For example, symmetry is an important concept in both fields. Physicists have used geometric ideas to model the world and the universe, and geometers have investigated physical problems.

Robotics

Computer vision is the ability of a robot's computer to recognize the shape and geometric features of an object before it can interact with the object, such as picking up a part from a manufacturing line to be used in the assembly of a larger component.

Robot motion planning is an issue in robot design. While the engineer and the planner know what they want the robot to do in a manufacturing or other type of process, the composition of the robot and the components used in its manufacture put restrictions on what movements it can actually perform. An understanding of this "movement space" and what can be

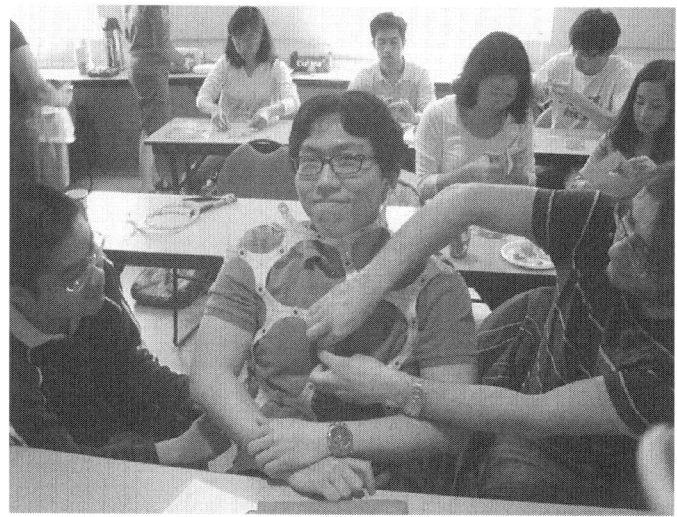 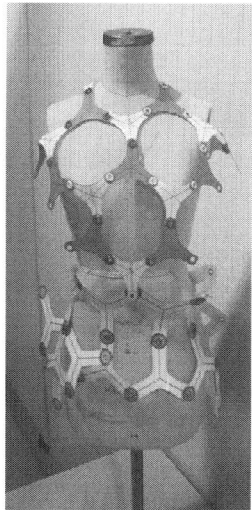

(Left) Graduate students in Cornell University's mathematics department try to make a tiling, which is fit to one student's body.

(Right) A sample tiling covering made by Professor Thurston. (both photos, Cornell University)

reached, held, moved, and so forth is a consideration of the geometry of the robot.

Geometry in Fashion Design

In March 2010, there was a fashion show of a Japanese fashion designer, Dai Fujiwara for Issey Miyake. It was not an ordinary fashion show but a place where fashion and advanced mathematics met. Dai Fujiwara was inspired by a legendary mathematician, William P. Thurston. Human bodies are beautiful geometric figures, which are curved in quite complicated ways. Covering these geometric objects with pieces of clothing in various types is certainly a place where mathematics can have a great influence.

A body is a surface of variable curvature. The top of the head, or shoulders, are positively curved parts, like spherical surfaces. The armpit is an example of a negatively curved part of the body, like a hyperboloid or saddle shape. Divide a circle into three arcs with equal-length and three points, A, B, and C, form endpoints of the arcs. Suppose there are hinges at A, B, and C so that the angle between two adjacent arcs can be changed. By changing the angle, one can make the deformed circle fit to a part of some surface. If the curvature is locally constant on some neighborhood of a point on the surface, and the size of the circle is small enough to be contained in that neighborhood, then this is possible. The curvature of the surface there should be same with the sum of all angle changes at the hinges. This idea was originally proposed by the great German mathematician, Carl Friedrich Gauss. A similar idea was proposed by Thurston. His idea was the following. Instead of a circle, consider Y-shape pieces. The three legs have the same length, and the angle between each pair of adjacent legs is 120 degrees. Suppose the size of the Y-shape piece is small enough. Connect endpoints of many Y-shape pieces by adding hinges, and let the hinges have some appropriate angles, and the result could fit on various types of surfaces. The angles that the hinges make determine the local curvature of the surface. If the surface is curved dramatically or the curvature of the surface is very large, then much smaller -shape pieces would be needed. This is one of the simplest ways to obtain a tessellation of a surface. Fashion designers have made use of these ideas in order to make beautiful coverings for the human body.

Geometry in Other Applications

In character recognition, a document is scanned and read on a computer; a computer is able to distinguish characters since they have certain configurations. If the image is clear, the recognition is simple. When the image is not clear, recognition becomes a much harder problem and geometry is brought to bear to try to differentiate characters. The algorithms used must be fast, however.

Social network theory involves the connections that people make in their social networks, which form a part of what can be studied using finite geometries. A 2009 survey of "friends" on Facebook showed that there was an average of 6.5 connections between any two randomly chosen participants.

Occupational Connections

Geometry is connected to a number of occupations and is used often in industry.

Carpenters, cabinetmakers, and construction managers are professionals who need to know, understand, and use the concepts of angle measurement, parallel lines, quadrilaterals, the Pythagorean Theorem (named for Pythagoras of Samos), area, and volume and need to know how to make and read three-dimensional drawings.

Surveyors, cartographers, photogrammetrists, and surveying technicians are professionals who need to know, understand, and use the concepts of angle measurement, congruent triangles, the triangle inequality, parallel lines, quadrilaterals, similarity, the Pythagorean Theorem, right-triangle trigonometry, circles, constructions, area, volume, and transformations and need to know how to make and read three-dimensional drawings.

Firefighters are professionals who need to know, understand, and use the concepts of area and volume.

Forest, conservation, and logging workers are professionals who need to know, understand, and use the concepts of angle measurement, congruent triangles, right-triangle trigonometry, area, and volume and need to know how and read three-dimensional drawings.

Automotive service technicians and mechanics are professionals who need to know, understand, and use the concepts of angle measurement, area, and volume.

Geometry has been useful in a wide variety of other professions also, including printing and the graphic arts, heavy equipment operation, fashion and apparel design, navigation, painting and paperhanging, engineering, home planning, plumbing and pipe fitting, outdoor advertising, landscape technology, and architecture and drafting, as well as optical technicians, machinists, cement workers, electricians, general contractors, and surveyors.

In the twenty-first century, geometry is connected to many occupations and fields within and outside mathematics. Students investigate geometric topics throughout their school experiences. Sometimes these experiences are in separate geometry courses, but often they are integrated with numerous mathematical perspectives and applications. In the nineteenth century, algebraist James Joseph Sylvester explained that

Time was when all the parts of the subject were dissevered, when algebra, geometry, and arithmetic either lived apart or kept up cold relations of acquaintance confined to occasional calls upon one another; but that is now at an end; they are drawn together and are constantly becoming more and more intimately related and connected by a thousand fresh ties, and we may confidently look forward to a time when they shall form but one body with one soul.

Further Reading

Dehaene, Stanislas, Véronique Izard, Pierre Pica, and Elizabeth Spelke. "Core Knowledge of Geometry in an Amazonian Indigene Group." *Science* 311 (2006).

Elam, Kimberly. *Geometry of Design: Studies in Proportion and Composition*. New York: Princeton Architectural Press, 2001.

Eppstein, David. "Geometry in Action." http://www.ics.uci.edu/~eppstein/geom.html.

Gibilisco, Stan. *Geometry Demystified*. New York: McGraw Hill, 2003.

Gorini, Cathy. *Geometry at Work: Papers in Applied Geometry*. Washington, DC: Mathematical Association of America, 2000.

Meyer, Walter. *Geometry and Its Applications*. 2nd ed. Burlington, MA: Elsevier Academic Press, 2006.

"Navigation: Using Geometry To Navigate Is Innate, At Least For Fish." *ScienceDaily,* August 15, 2007. http://www.sciencedaily.com/releases/2007/08/070813121027.htm.

Pierro, Mike, et al. "Geometry: Career Related Units. Teacher's Edition." Minnesota State Department of Education, 1973. http://www.eric.ed.gov/PDFS/ED085548.pdf.

Sinclair, Nathalie. *The History of the Geometry Curriculum in the United States*. Charlotte, NC: Information Age Publishing, 2008.

Whiteley, Walter. "The Decline and Rise of Geometry in 20th Century North America." In *Canadian Mathematics Study Group Conference Proceedings*. Edited by J. G. McLoughlin. St John's: Memorial University of Newfoundland, 1999. http://www.math.yorku.ca/Who/Faculty/Whiteley/cmesg.pdf.

David C. Royster
Hyungryul Baik

Geometry of Music

Category: Arts, Music, and Entertainment.
Fields of Study: Communication; Connections; Geometry; Representations.
Summary: The mathematical principles of symmetry and scaling play important roles in musical composition.

Musical information can often be represented naturally with shapes, allowing insights to be gained from geometric techniques.

One indication of the close connection between music and geometry comes from the fact that Euclid of Alexandria, who wrote *Elements of Geometry* (300 B.C.E.), a founding document of geometry, also wrote a comprehensive treatise on the mathematics of musical pitches, *Theory of Intervals*. The eighteenth-century mathematician Leonhard Euler also developed geometric tools for music analysis.

Symmetry is one of the most powerful ideas in geometry. No less so in the geometry of music, where symmetries abound. Geometric techniques can be applied to musical scales, chords, and melodic lines. Because of the concept of octave equivalence, the 12 pitches of the equally tempered chromatic scale are inherently cyclic in nature. Thus, the geometric theory of cyclic groups plays a major role in the mathematical description of scales and chords. Similarly, geometry can play a role in the analysis of musical rhythm, particularly in musical forms based upon a repeating rhythmic motif.

In twentieth-century atonal music, geometric ideas have been proposed as unifying theoretic structures to fill the role once played by tonal harmonic concepts.

Symmetries in the Twelve Pitches of the Equally Tempered Scale

Two fundamental principles of modern musical analysis are "octave equivalence" and "equal temperament." Octave equivalence refers to the perception, believed to be universal in developed music cultures, that two pitches separated by an octave are members of the same "pitch class." Equal temperament refers to the system of musical intonation by which the 12 chromatic half steps within the octave represent uniform frequency scaling—given a pitch with frequency f, the pitch one half step above has frequency $2^{1/12}f$. In the equally tempered scale, enharmonically spelled notes, such as C♯ and D♭, represent the same pitch.

The twelve pitch classes are inherently cyclic. This principle is represented in the left view of Figure 1, which is identical to an analog clock face, with the traditional "12" replaced by "0." The diatonic scale is represented by the vertices of the inscribed polygon in the center view of Figure 1. This arrangement of the seven diatonic pitches is the most even spacing possible for seven pitches in the 12-tone octave. The evident symmetry about the 2–8 axis puts the complicated diatonic sequence of half steps and whole steps into a simpler conceptual framework. The figure illustrates that the Dorian Mode (which begins and ends on the second diatonic scale degree, given here as "D" or "2") is unique

Figure 1. The 12 pitch classes.

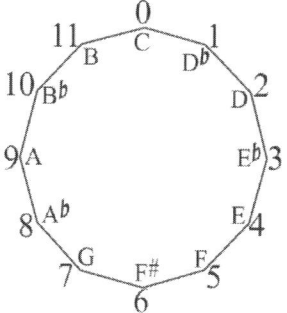 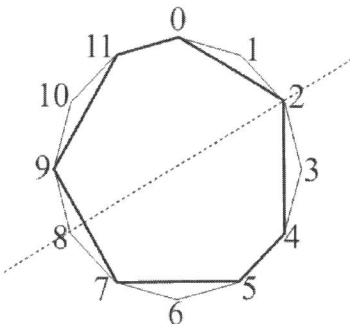 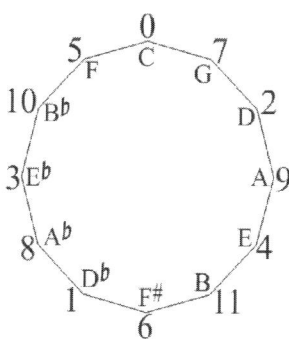

Left: The 12 pitches of the equally tempered chromatic scale arranged on a circle. Center: The vertices of the inscribed polygon represent the pitches of the diatonic scale. The diatonic arrangement is the most evenly spaced distribution of seven vertices in a 12-sided figure. Note the symmetry inherent in the Dorian Mode, which begins and ends on pitch 2 (D). Right: Diametric reflection of the odd-numbered pitches results in the circle of fifths.

Figure 2. The first eight rows of a Tonnetz (or Tone Network).

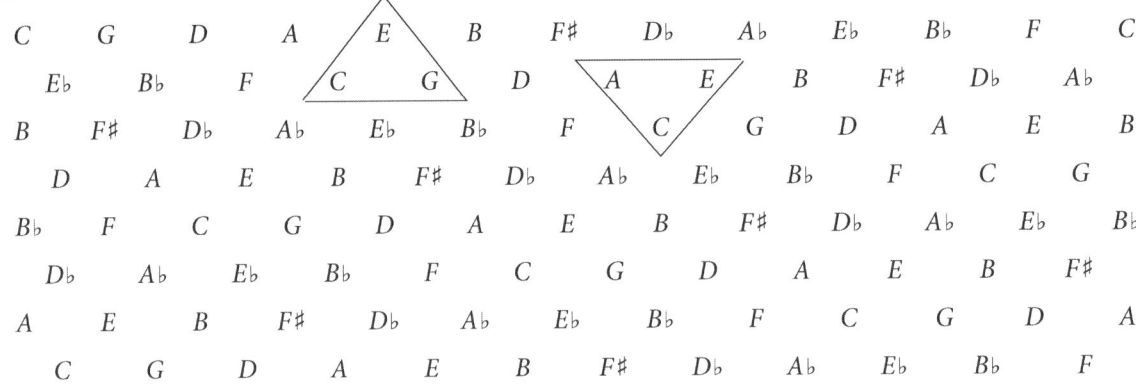

The pitch classes of the circle of fifths are arranged horizontally. The vertical alignment of the pitch classes is chromatic. Diagonals in the southeast direction progress by intervals of the minor third. Northeast diagonals progress by major thirds. All tonal sonorities are given in this representation by polygons containing adjacent pitches. For example, major triads are given by triangles with vertex at top and minor triads are given by triangles with a vertex at the bottom, as shown above for the C major and A minor triads.

among the diatonic modes in that it follows the same sequence of intervals both ascending and descending.

The six pairs of diametrically opposite pitch classes in the clock representation are separated by the interval of a "tritone," so named because it contains three whole steps. In tonal music, the tritone is considered the most dissonant-sounding interval. If the three odd-numbered pitch class pairs on the clock face are reflected diametrically, the result is the "circle of fifths" shown in the right view of Figure 1. The circle of fifths is familiar to music students as a mnemonic device for learning the musical key signatures: the number of sharps increases by one (or alternatively, the number of flats decreases by one) at each step in the clockwise direction, while the number of flats increases (or sharps increase) at each step in the counterclockwise direction. The circle of fifths is used extensively as an analytical tool for twentieth-century music in the work of American composer and music theorist Howard Hanson.

Representing Musical Structure in Geometric Spaces

Beginning with the musical writings of Euler and continuing at least through the work of the influential

Figure 3. Eighth-note subdivisions of rhythmic units arranged around a circle.

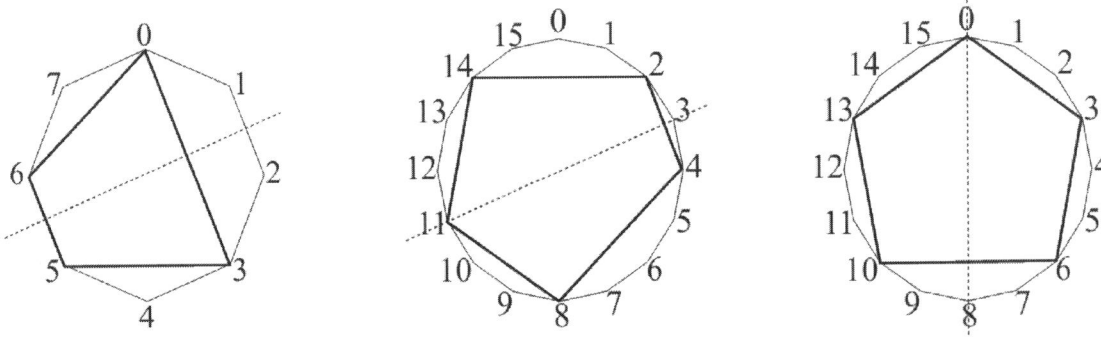

Left: The vertices of the inscribed polygon represent the handclap rhythm heard in the Elvis Presley recording of "Hound Dog." Center: The vertices of the inscribed polygon represent the well-known clave rhythm heard in Afro-Cuban music. Right: The bossa nova cowbell rhythm heard in Quincy Jones's "Soul Bossa Nova."

music theorist Hugo Riemann (not to be confused with the mathematician Bernhard Riemann) in the nineteenth century, the representation of harmonic concepts in a two-dimensional array called a "Tonnetz" (Tonal Network) has guided the understanding of tonal harmony. In the tonnetz shown in Figure 2, the rows are simply the entries of the circle of fifths, while the columns are the 12 diatonic pitch classes arranged chromatically (by half steps). The result is that the diagonals are made up of pitch classes separated by minor thirds (in the southeast direction) and major thirds (in the northeast direction). In this arrangement, the sonorities of tonal harmony can be represented by polygonal groupings of the adjacent symbols: triangles for major and minor triads, parallelograms for major and minor seventh chords, and similar structures for diminished, augmented, and dominant seventh chords. The musical theory of "modulation" (changing from one tonal center to another in the course of a musical composition) is aided by the geometric perspective of a Tonnetz. Tonal networks such as the one shown here are precursors of the contemporary musical theory of "pitch class spaces."

Recently, chords have been modeled as points in geometric spaces called "orbifolds." Music theorists analyze the symmetry of chords inside of the space with respect to translation, reflection, or permutation and look at short line segments between structurally similar chords.

Rhythmic Symmetry

Like the 12 pitch classes, the metrical organization of music in time is also highly cyclic, allowing similar geometric techniques to be applied to rhythm. The left view of Figure 3 shows the eighth-note subdivisions of a 4/4 measure. The vertices of the inscribed polygon represent the rhythmic placement within the measure of the handclap rhythm from the iconic 1956 Elvis Presley recording of "Hound Dog." This complicated rhythm has a simple symmetric structure when viewed geometrically. Similarly, the center view in Figure 3 shows the clave rhythm familiar to listeners of Afro-Cuban music, with its line of symmetry. The left view of Figure 3 shows a characteristic bossa nova rhythm (which can be heard on the cowbell in Quincy Jones's "Soul Bossa Nova") and its line of symmetry.

Further Reading

Archibald, R. C. "Mathematicians and Music." *American Mathematical Monthly* 31, no. 1 (1924).

Demaine, E. D., F. Gomez-Martin, H. Meijer, D. Rappaport, P. Taslakian, G. T. Toussaint, T. Winograd, and D. R. Wood. "The Distance Geometry of Music." *Computational Geometry: Theory and Applications* 42, no. 5 (2009).

Hall, Rachel Wells. "Geometrical Music Theory." *Science* 320 (2008).

Johnson, Timothy. *Foundations of Diatonic Theory: A Mathematically Based Approach to Music Fundamentals.* Lanham, MD: Scarecrow Press, 2008.

Eric Barth

Golden Ratio

Category: Arts, Music, and Entertainment.
Fields of Study: Measurement; Number and Operations; Representations.
Summary: The golden ratio of roughly 1.618 is found throughout nature and art.

It was Euclid of Alexandria, a well-known Greek mathematician, who in his book *The Elements* (300 B.C.E.) first wrote about the golden ratio. The golden ratio is denoted by the Greek letter φ (phi) and known also as the "golden section," the "golden mean," and the "divine proportion."

This last name was given to φ because of the frequency with which the ratio exists in the natural world—leading many to hold it up as a mystical number. The golden ratio is, as all ratios are, a comparison. In his description, Euclid describes the golden ratio through the division of a line segment. A line segment whose length is A is divided into two smaller pieces, one of length B and the other of length C, such that the ratio of the original segment to the larger piece is equal to the ratio of the larger piece to the smaller piece. Mathematically, this ratio would be represented as the following:

$$\frac{A}{B} = \frac{B}{C}.$$

A perfect rectangle is a rectangle in which the ratio of the length of the longer sides to the length of the shorter sides yields φ. Alternatively, the ratio may be expressed as follows:

$$\frac{1+\sqrt{5}}{2}$$

and it is approximately equal to 1.16180339877. . . . As this is an irrational number, there is no end to its digits and no pattern among them.

The golden ratio may be used to create a golden spiral. Golden spirals are common in nature and can be found on shells, the caverns of the inner ear, the horns of various animals, and even some flowering plants. A golden spiral is a spiral that gets wider by a factor of φ for every quarter turn it takes as it opens outward from the point of origin (see Figures 1–2). If one considers the origin to be the eye of a hurricane, the spiraling out can be seen in the shape of the hurricane (the circling of winds that opens outward from the eye), and this provides yet another example of the golden ratio's appearance in nature.

The golden ratio appears in many other areas as well, including science, art, and nature. For example, the work of Herodotus (fifth century B.C.E.), considered the first historian, indicates the use of the golden ratio in the construction of the pyramids (see Figure 3). Phiddias (490–430 B.C.E.), a sculptor, is said to have used the golden ratio in the creation of sculptures that were later found in the Parthenon. The Parthenon itself consists of many uses of the golden ratio, a simple example being the length and width of the building. Similarly, the golden ratio appears in modern architecture, such as the United Nations Building in New York City. Here the ratio of the height of every 10 floors as compared to the width of every 10 floors also yields the golden ratio.

The work of Leonardo da Vinci is also said to incorporate the golden ratio, including in the definition of the proportions in the Mona Lisa.

The use of the golden ratio in art and architecture is common, especially when one considers that the ratio is pleasing to the eye. Gustav Fechner (1801–1887) performed many experiments with respect to this ratio. He found that rectangles, books, buildings, and other objects were more pleasing to individuals when they contained the golden ratio.

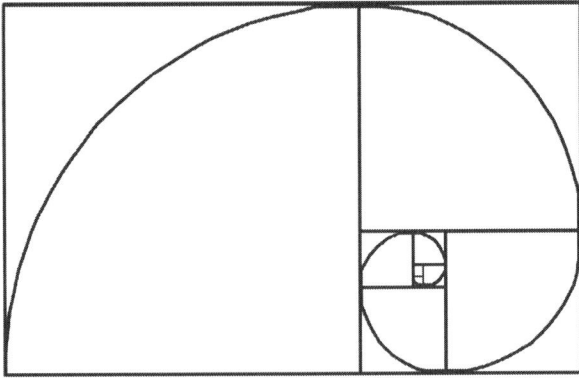

Figure 1. A golden spiral.

Figure 2. A golden spiral in a seashell. (Photos.com)

Music is another place where the golden ratio plays a vital role. Mozart's piano sonatas use the golden ratio in the arrangement of sections of measures that make up individual pieces. Mozart's piano sonatas are made up of two sections called the "exposition" and the "recapitulation." In one 100-measure composition, Mozart divided the pieces into two sections between the 38th and the 62nd measures. The measures in the pieces, when compared, yield the closest approximation to the golden ratio that can be made when dividing a 100-measure composition into two sections. However, the pieces do not always make use of the golden ratio throughout. That is, subsections do not always include the golden ratio, leading some to question whether Mozart was conscious of his use of it. In addition, in many of the most successful musical pieces, the climax of the piece occurs in accordance with φ. That is, the ratio between the length of the piece prior to the climax compared to that after the climax yields, once more, the golden ratio.

Figure 3. The golden ratio in pyramids.

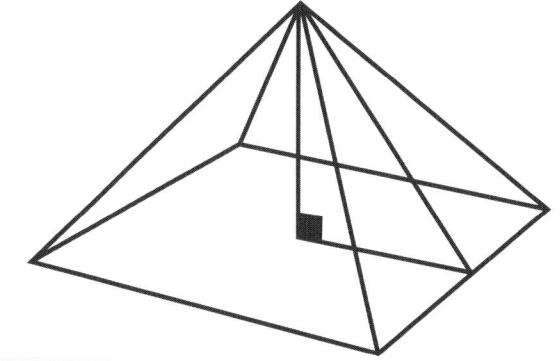

(Photos.com)

Further, the golden ratio is apparent in proportions in the human body. If the distance from the navel to a person's foot is considered to be "1," then the height of the person is approximately φ. The ratio of the distance from the navel to the top of the head to the length of the head also approximates φ. In the idealized human face (that which is said to be most beautiful in terms of proportions, φ comes up when comparing the length of the face to the width; the length of the mouth and the width of the nose, and many other comparisons.

The golden ratio is also related to the Fibonacci sequence—a numeric sequence in which each successive term (except for the first two) is obtained by adding the two prior terms. This yields 1, 1, 2, 3, 5, 8, 13, 21, ... When the ratios between successive terms in the sequence are found, they approach the golden ratio.

Some question whether the golden mean is a number that is preferred or significant in nature or whether the number is so prevalent because the mathematical meaning of the number influences or biases perceptions of the applicability. The diversity of systems in which it appears, including multiple developmental markers of human growth, suggests that it may be broadly advantageous. Analysis shows that the ratio's logarithmic spiral is a system that could theoretically self-replicate indefinitely. It also minimizes wasted space and gives new growth maximum exposure to necessary resources, such as sunlight. This makes a golden spiral an optimal and efficient design for growth in biological systems.

Further Reading

Dunlap, R. *The Golden Ratio and Fibonacci Numbers*. London: World Scientific Publishing, 1998.

Hemenway, P. *Divine Proportion: Phi in Art, Nature and Science*. Salt Lake City, UT: Sterling Press, 2005.

Livio, M. *The Golden Ratio: The Story of Phi, The World's Most Astonishing Number*. New York: Broadway Books, 2002.

Lidia Gonzalez

Gymnastics

Category: Games, Sport, and Recreation.
Fields of Study: Algebra; Data Analysis and Probability; Geometry.
Summary: Performing gymnastics depends upon an understanding of geometry and forces.

Gymnastics is an athletic performance activity that depends on balance, flexibility, and strength for producing graceful movements. Gymnastics can be recreational or competitive. There are also numerous forms of gymnastics, including artistic, acrobatic, and aerobic. The main mathematical topics involved in gymnastics include the mechanics of motion, patterns in choreography, and competition scoring systems. Mathematics has sometimes been described as "mental gymnastics."

Rotations

Many gymnastics routines include rotations. The key mathematical characteristic of a rotating body is its angular momentum, which is equal to the product of the mass, the velocity, and the distance between the center of mass and the axis of rotation. When there are no external forces, the angular momentum is conserved—it does not change. Gymnasts cannot change their mass, but they can reposition their center of mass relative to the axis of rotation, making the speed change to preserve the momentum. When a rotating gymnast tucks in closer to the center of rotation, the speed increases. For example, a gymnast can hold onto a bar by the hands and keep the body straight, making a relatively slow rotation around the top uneven bar called "giant swing." As the gymnast tucks his or her limbs in closer to the bar, the center of mass becomes closer to the axis of rotation, and the gymnast spins faster. Mathematics also helps determine the optimal angle at which the gymnast should release from the bar in order to perform subsequent transitions and maneuvers.

While simpler routines can be performed intuitively, through trial and error, competitive gymnasts develop complex sequences of moves that involve detailed calculations of mass, momentum, velocity, position of the apparatuses, and so on. Conversions between rotation and moving along straight lines are a part of many routines, with speeds and directions determined by conservation of momentum laws. For example, a gymnast runs to a springboard, accumulating momentum. As the gymnast jumps off the springboard, the vectors of the momentum generated by the springs and the momentum of the run are added together, propelling the gymnast forward at about a 45-degree angle to the floor. The gymnast can then push off a horse ahead, converting momentum into the angular momentum of rotating the body around the horse. At the highest point of this rotation, the gymnast can tuck the limbs in, moving the mass close to the axis of rotation and accelerating for a flip in the air. For example, a triple back somersault involves two and three-quarter body rotations before landing. Before landing, the gymnast straightens out, moving the limbs farther from the axis of rotation and slowing down the rotation, allowing for a soft, safe landing on the feet.

Scoring of Artistic Gymnastics Competitions

The current system of scoring in artistic gymnastics is relatively complex. It assigns a difficulty score to the attempted routine and then subtracts from that score for mistakes in execution. The score is analytic—it is based on decomposing gymnastic routines into individual elements. Existing elements are summarized in the illustrated Table of Elements, and given difficulty ratings from A (0.1 points) to G (0.7 points). Additions to the Table of Elements are frequently named after gymnasts who first performed them successfully. Such new elements are submitted by the competing gymnasts ahead of the competition event, to be evaluated by an international committee.

Eight highest difficulty values of the routine, added together, form the difficulty value (DV). Skills from five required Element Groups are awarded 0.5 points each, for the maximum 2.5 points in composition requirement (CR). Finally, an additional 0.1 or 0.2 points are given for each element if elements are connected, which adds to connection value (CV). The difficulty score (D) is the sum of these points: $D = DV + CR + CV$.

In addition to the difficulty score, there is an evaluation of the artistry and execution called "E-score." The judges take away points from the perfect 10.0 E-score for technical or artistry mistakes. Each fall costs 1 point.

Trampolining and Conservation of Energy

Many gymnastic apparatuses are somewhat springy. Trampolining is a type of gymnastics that occurs entirely on trampolines and uses flight-like moments between contacts with the surface for striking routines. Trampolining involves the accumulation of energy. First, the kinetic energy of the gymnast's limb flexes and motions is converted into the potential energy of the stretched trampoline fabric. Then, the gymnast is thrown in the air, converting this potential energy into the kinetic energy of the motion. As the gymnast gains height, the kinetic energy is converted into the potential energy again. Gravity pulls the gymnast down with acceleration, converting to kinetic energy, which converts to the potential energy of the stretched trampoline upon contact, and so on.

From the point of view of mechanics, the trampoline is a device for storing the gymnast's potential energy between jumps. This view can explain, for example, why gymnasts cannot jump infinitely high, adding more and more energy to the trampoline. The maximum stretch of the trampoline limits the amount of energy stored in it. This can also be used to compute the theoretical maximum height of a trampoline jump.

Different types of gymnastics are easier to perform with different body types. A lower body-mass-to-height ratio makes it easier to twist during movements and to hide momentum transitions in the twisting, so tall, skinny people are better suited for artistic gymnastics. In trampolining, twists and transitions are not as crucial as higher rotation speeds and are easier with a higher body-mass-to-height ratio. Also, both take-offs and landings on trampolines require significant bursts of energy and muscle strength. Therefore, shorter, stockier athletes are better suited for trampolining.

Further Reading

Jemni, Monem, ed. *The Science of Gymnastics*. New York: Routledge, 2011.

Roper, Tom. "Mathematics and the Motion of the Human Body." *The Mathematical Gazette* 74, no. 467 (1990).

———. "Mathematics and the Motion of the Human Body, Continued." *The Mathematical Gazette* 74, no. 468 (1990).

Sommer, Christopher. *Building the Gymnastic Body: The Science of Gymnastics Strength Training*. Mesa, AZ: Olympic Bodies, 2008.

Maria Droujkova

Harmonics

Category: Arts, Music, and Entertainment.
Fields of Study: Algebra; Measurement; Number and Operations; Representations.
Summary: Harmonics are sonic components that are periodic at and integer multiples of the fundamental frequency.

Harmonics are components of a musical sound with well-defined frequency relationships to one another. For a pitch of frequency f, typically measured in units of cycles per second, or hertz (Hz), the nth harmonic has frequency $n \times f$. In this context the frequency f is referred to as the "fundamental frequency." Harmonics are closely related to overtones (or partials), which are defined to be secondary pitches that audibly resonate when a fundamental pitch sounds. The number and strength of these secondary pitches are responsible for the distinct timbres perceived in different instruments or voices. The overtone series in music (also called the harmonic series at the risk of confusion with the infinite sum of the same name) refers to the sequence of ascending harmonics with frequencies $2f$, $3f$, $4f$. . . . With only a few exceptions, the pitches of the lower harmonics match well with the frequencies of 12 pitches of the equally tempered scale. Further along the overtone series, the pitch spacing becomes very small—smaller than the traditional half step—and these upper harmonics, if heard, would sound distinctly out of tune. With the discovery of the overtone series by Jean-Philippe Rameau in the eighteenth century, the notion of musical consonance as the exclusive natural and rational sonic phenomenon—pursued by mathematicians from Pythagorus of Samos to Leonhard Euler—began to fade. There is a close physical relationship between the harmonic frequencies and the length of the vibrating medium. This relationship is exploited in the performance practices of musical instruments.

Vibrating Media and the Overtone Series

For vibrating strings (such as violins and guitars) and open vibrating air columns (such as the Western concert flute and some organ pipes), the words "harmonic," "partial," and "overtone" are essentially synonymous, with a slight difference in the enumeration: the fundamental pitch (frequency f) is referred to as the "first harmonic." The first overtone (frequency $2f$) refers to the second harmonic, and so on. In stopped air columns (such as the clarinet and some organ pipes), the overtone series omits certain harmonic frequencies. For vibrating membranes (such as percussion instruments), overtones may exist at nonharmonic frequencies.

It is therefore a slight abuse of terminology to refer, as is commonly done, to the sequence of harmonics as the "overtone series." Physically, the overtone series is seen by observing the motion of a vibrating string of length L and natural frequency f. If forced to vibrate at frequencies $n \times f$ (for $n = 2, 3, \ldots$), $n-1$ stationary points (nodes) appear along the string, at intervals of L/n. In effect, the string moves as n strings of length L/n joined end to end. String performers utilize this fact by lightly stopping the string at lengths $L/2, L/3, \ldots L/n$ to produce flute-like harmonic tones (sometimes called "flageolet tones").

From the overtone perspective, only lower harmonics are perceptible to the hearer of a fundamental pitch. The first six harmonics are perceived by the modern hearer as in tune within the 12 pitches of the equally tempered scale, in which the octave (the distance between the first and second harmonic) is divided into 12 equal half-step intervals. The frequency difference between successive pitches in this 12-tone system is given by

$$f_{n+1} = 2^{1/12} f_n.$$

The second, fourth, and eighth harmonics, at octaves above the fundamental, sound perfectly in tune. Upper harmonics can sound significantly out of tune, however. The seventh harmonic sounds uncomfortably flat compared to its nearest corresponding equal temperament pitch. The 11th harmonic has a frequency almost equidistant between adjacent notes of the equally tempered scale, causing it to sound very out of tune—likewise for the 13th and 14th harmonics.

These considerations are significant for period-instrument brass performers, whose instruments, like the so-called "natural trumpet," are nothing more than long tubes without the length-changing system of valves of modern trumpets. Performers play tunes on these instruments by producing overtones, typically between the 3rd and 16th in the series.

While skillful performers can compensate for the most problematic overtones, composers in the baroque era typically avoided these notes or used their sonic character for special effect. Modern composers have experimented with specially tuned pianos and electronic instruments to directly explore the sonorities of harmonics. The first 24 harmonics are listed in Table 1 with fundamental pitch taken as the A below middle C. Harmonics with frequencies that differ significantly from the equally tempered scale are indicated in bold type.

Other Uses of the Word "Harmonic" in Mathematics

In mathematics, the word "harmonic" appears in a number of contexts, all of which trace their origins to the overtone series and associated physical vibrations. A harmonic progression is defined as the term-by-term reciprocal of an arithmetic progression. For example, the arithmetic sequence $a_1 = 1, a_2 = 2, a_3 = 3, \ldots, a_n = n$ gives rise to the harmonic sequence $h_1 = 1, h_2 = 1/2, h_3 = 1/3, \ldots$, where $h_n = 1/n$. In this example, the arithmetic sequence gives the frequency multiples for the overtone series, and the harmonic sequence corresponds to the wavelengths of the respective overtones. The harmonic mean is the reciprocal of the arithmetic mean of reciprocals.

For example, the harmonic mean of two numbers x and y is defined as $2(1/x + 1/y)^{-1}$. The harmonic series in mathematics is the infinite sum $1 + 1/2 + 1/3 + \ldots$,

Table 1. The first 24 harmonics of a selected fundamental frequency. Also listed is the nearest pitch in the equally tempered scale. Note that some upper harmonics deviate substantially from pitches of the 12-tone scale.

Nearest Pitch (Hz)	Frequency (Hz)	Harmonic	Nearest Pitch (Hz)	Frequency (Hz)	Harmonic
A (220)	220	1st harmonic	F♯ (2960)	**2860**	**13th harmonic**
A (440)	440	2nd harmonic	G (3136)	**3080**	**14th harmonic**
E (659)	660	3rd harmonic	G♯ (3322)	3300	15th harmonic
A (880)	880	4th harmonic	A (3520)	3520	16th harmonic
C♯ (1109)	1100	5th harmonic	A♯ (3729)	3740	17th harmonic
E (1318)	1320	6th harmonic	B (3951)	3960	18th harmonic
G (1568)	**1540**	**7th harmonic**	C (4186)	4180	19th harmonic
A (1760)	1760	8th harmonic	C♯ (4435)	4400	20th harmonic
B (1976)	1980	9th harmonic	**D (4698)**	**4620**	**21st harmonic**
C♯ (2218)	2200	10th harmonic	**D♯ (4978)**	**4840**	**22nd harmonic**
D (2349) / D♯ (2489)	2420	11th harmonic	E (5274)	5060	23rd harmonic
E (2636)	2640	12th harmonic	E (5274)	5280	24th harmonic

providing the canonical example of a series whose terms approach zero, but nevertheless, the sum diverges. The harmonic oscillator is a differential equation whose solutions are sinusoidal functions that can be used to model musical sounds. Harmonic analysis is the study of functions (or signals) by decomposition into fundamental component functions by means of the Fourier transform or other techniques. In the study of complex variables, harmonic functions are generalizations of the sinusoidal functions that model fundamental vibrations.

Further Reading

Cohen, H. F. *Quantifying Music: The Science of Music at the First Stage of the Scientific Revolution, 1580–1650*. Dordrecht, Netherlands: D. Reidel Publishing, 1984.

Gouk, Penelope. "The Role of Harmonics in the Scientific Revolution." In *The Cambridge History of Western Music Theory*. T. Christensen, ed. Cambridge, England: Cambridge University Press, 2002.

Johnston, Ben. *Suite for Microtonal Piano*. Robert Miller, piano; New World Records, 80203-2.

Sundberg, Johan. *The Science of Musical Sounds*. San Diego, CA: Academic Press, 1991.

Eric Barth

(Photos.com)

Hockey Equipment

Hockey equipment also benefits from mathematics. Helmets have become mandatory in most hockey leagues, and researchers are continually seeking ways to better disperse the powerful kinetic energy of blows and collisions. Iconic player Robert "Bobby" Hull is credited with introducing curved blades on hockey sticks, which improves control and accuracy. Many players still use traditional wooden hockey sticks, but researchers have also developed flexible, lightweight composites and aluminum sticks, often involving statistical analyses and modeling. Physicists have also used mathematical models to analyze the characteristics of different hockey shots.

Hockey

Category: Games, Sport, and Recreation.
Fields of Study: Data Analysis and Probability; Geometry; Measurement.
Summary: Playing hockey is an application of geometry, as players in constant motion determine angles of approach, plot routes through opponents, and visualize the vector of the puck.

Ice hockey is a team sport played on an ice rink by skating players using sticks to move a rubber disk called a "puck" into the opposing team's goal. Field hockey and street hockey are usually played on foot, either on grass fields or street surfaces, using a ball. There is evidence that hockey-style games have existed for millennia, and ice hockey has long been popular in parts of the world that are cold enough for long-lasting seasonal ice. The basic rules of modern ice hockey were developed in Canada in the late 1800s, and the National Hockey League of North America (NHL) dates back to the early 1900s. The growing prevalence of indoor ice rinks has allowed hockey to expand into warmer places, like Florida and California, with mixed success. Ice hockey is highly geometric, in terms of both player action and the surface on which it is played. Mathematics and statistics are also used to model various aspects of game play and to develop improved equipment.

Geometry

A hockey rink is in some ways more geometric than other sports surfaces. Overall, the ice is essentially rectangular. North American professional rinks have corners that are rounded on a circle with a radius of 28 feet. Rinks have mirror symmetry end-to-end and side-to-side, including five circles used for face-offs. The goalie primarily occupies the space in front of the goal known as the "crease," which is a half-circle with a six-foot radius in international play. In North American professional rinks, the crease is truncated to eight feet wide by transecting lines drawn one foot on either side of the six-foot-wide goal. Aside from the crease, goalies in some professional leagues may play the puck only in the goaltender's trapezoid. This symmetrical region has one 18-foot base formed by the goal line and another 28-foot base determined by the boards (the wall behind the goal).

Hockey also requires an awareness of geometry for competitive play. Players are in constant motion and thus always calculating the best angle at which to approach an opponent, based on the opponent's speed and trajectory, as well as the best route through the moving players. Turning and stopping on ice require different applications of forces than sports played on foot, with arcing turns or various radiuses being more common than point pivots and sudden reversals. Being a hockey goalie is an ongoing exercise in mathematics and physics. Geometric ideas like circumferences, radiuses, and angles are very important, as is the ability to visualize vectors. Goalies shift within the crease in response to the continuously changing locations of other players in the plane of the rink to simultaneously minimize opponents' possible angles of attack and maximize their ability to intercept the puck. Time series analyses of several decades of data have shown that NHL games steadily average about 30 shots on goal per 60-minute game. There have been vocal critics of the artificial intelligence used for hockey goalies in some video games, with assertions that the programming fails to accurately mimic the sort of continuous precision adjustments used by real goalies. Hockey terminology has been used with some students to motivate and teach geometric concepts.

Statistics

Sports fans have become increasingly interested in studying sports statistics for prediction and deeper analyses. Operations researchers Jack Brimberg and William Hurley investigated the common belief that the first goal in the game "sets the tone" for the rest of the game. They calculated that the team that scored first was more likely to win, especially if the first goal was scored later in the game. Others have analyzed the way in which the NHL determines which teams will compete in the play-offs. There are 82 games in the regular NHL season. Points are awarded to the teams as follows: two points for winning the game, zero points for losing in a regulation 60-minute game, but one point for losing if the game went to overtime. No other league rewards a team differentially for losing in overtime. The intent is purportedly to keep tied teams playing competitively in the third period. However, data suggest that teams tend to rein in play and allow the game to go into overtime, which mathematical game theory suggests is the better move, because the reward for winning is the same, but the penalty for losing is reduced. A European system changes optimal strategy because the winner gets only two points in overtime versus three.

Other Connections to Mathematics

In climate science, Michael Mann, Raymond Bradley, and Malcolm Hughes quantitatively reconstructed temperature trends for the last 1000 years, producing a controversial graph called the "hockey stick graph," since its changes in slope resemble the bend of a hockey stick. One theorem regarding diagonals in Pascal's Triangle, named for Blaise Pascal, is also sometimes known as the "hockey stick theorem" for the shape it produces.

Further Reading

Brimberg, Jack and W. J. Hurley. "A Note on the Importance of the First Goal in a National Hockey League Game." *International Journal of Operational Research* 6, no. 2 (2009).

Gill, Paramjit. "Late-Game Reversals in Professional Basketball, Football, and Hockey." *The American Statistician* 54, no. 2 (2000).

Hache, Alain. *The Physics of Hockey*. Baltimore, MD: Johns Hopkins University Press, 2002.

BILL KTE'PI

Knots

Category: Games, Sport, and Recreation.
Fields of Study: Geometry; Representations.
Summary: Mathematical knots are useful in physics and biochemistry.

Since ancient times, knots have been used in sailing, building, textiles ("knit" comes from "knot"), climbing, and in recreation, as well as serving as symbols for spiritual or religious concepts like eternity or wisdom. Topology generalizes the idea of a knot to an embedded circle in 3-dimensional Euclidean space. In knot theory, a knot is a tangled-up loop, like a piece of string with the ends fused together. The simplest is the unknot, simply an untangled loop like a rubber band. Two knots are the same if one can be manipulated (transformed) into the other without breaking the loop or passing the string through itself. In 1926, Kurt Reidemeister demonstrated that all such transformations were made up of a sequence of just three basic moves called Reidemeister moves. Deciding whether two knots are the same via a sequence of such moves is a member of a host of problems involving changing one object into another without breaking or tearing, which have long stumped topologists. Topologists find it difficult to assure themselves that failing to transform one knot into another truly reflects impossibility, or rather just their own failure. In modern times, mathematical knots are useful in physics and biochemistry.

Invariants and Links

To wrestle with this problem, topologists have created an assortment of invariants, mathematical entities that can be unambiguously computed for each knot. If a particular invariant has different values on two knots, then those knots are different. Unfortunately, different knots can have the same invariants. In 1928, James Waddell Alexander II created a method for associating a polynomial to a knot, now called its Alexander polynomial. In 1983, Vaughan Jones, studying a simplified model of phase transitions, such as freezing, discovered a second invariant, the "Jones polynomial." Another mathematician, Edward Witten, soon noticed that the same polynomial could be computed from an invariant on particular three-dimensional spheres, providing insight into another difficult classification problem. Witten and Jones shared part of the Field's Medal in 1990 for these discoveries. Victor Vassiliev has since created a host of new invariants. The Vassiliev invariants are infinite in number, and it is conjectured that any two different knots will differ in at least one such invariant.

Not all invariants are polynomials. Henri Poincaré created a topological invariant called the "fundamental group." Applied to knots, it is called the "knot group" and is actually computed on the complement of the knot, that is, the abstract concept of all space with the knot removed. Poincaré's invariant was the seed of an area that grew into a central focus of twentieth-century mathematics called "homological algebra."

Knots, and their close cousins, links, have proven useful in a branch of physics called "topological quantum field theory." For this application, physicists use particular guidelines to trace knots in two dimensions. The knot diagrams then portray scenarios in which particles are created, interact, and are finally annihilated. By appropriately labeling pieces of knots, mathematicians can realize the Jones and other invariants via important modern mathematical constructs, including the Yang–Baxter equations and quantum groups. Mikhail Khovanov has created a new type of invariant on links, keeping this topic at the very forefront of contemporary mathematics.

Applications in Biochemistry

The application of knot theory to DNA molecules has helped to elucidate their biochemistry. The DNA molecule of a bacterium closes into a circle, which bends and twists itself into a knot. This knotted structure can block DNA replication. Using electron microscopy or gel electrophoresis, the biologist can determine an individual molecule's crossing and unknotting numbers, two numbers that classify knots. Enzymes called "topoisomerases" release the knots as a preliminary step to DNA replication. By carefully examining the knots that arise, molecular biologists have determined that there are two different topoisomerase molecules. Topoisomerase I releases the knot by cutting both strands of the molecule, and Topoisomerase II nicks just one strand and twists the cut strand around the other.

Further Reading

Adams, Colin C. *The Knot Book: An Elementary Introduction to the Mathematical Theory of Knots*. Providence, RI: American Mathematical Society, 2004.

Menasco, W., and L. Rudolph. "How Hard Is It to Untie a Knot?" *American Scientist* 83 (1995).

Sossinsky, Alexei. *Knots: Mathematics With a Twist.* Cambridge, MA: Harvard University Press, 2002.

Michael Klucznik

Lotteries

Category: Games, Sport, and Recreation.
Fields of Study: Data Analysis and Probability; Number and Operations.
Summary: A successful lottery depends on assuring the randomness of its selections and maintaining the perception of fairness.

Lotteries, which can be thought of as games that involve a winner selected by chance, have played an important role in the development of societies for more than 2000 years. Lotteries can include those run by political bodies, like states, where the winnings are money, or those run by a sports entity, like the National Basketball Association (NBA) Draft Lottery, where teams get to select new members. The U.S. government runs a Green Card Lottery program and selects winners using a computer-generated drawing. In most lotteries, very few people win anything substantial, and the purchase of a lottery ticket usually amounts to an unfair bet, in that the price of a single ticket is less than the average payoff across all tickets.

Nevertheless, lotteries are quite popular and consequently can raise substantial funds or allocate a small number of goods, services, or sought-after players among a large number of people or teams. The mathematical concepts of "randomness" and "expected value" are fundamental to the operation of lotteries and perceptions of fairness. Probability methods, especially combinations and permutations, are used to compute the odds or chances of winning, given certain conditions.

Distribution of Winnings

If lottery commissions somehow redistributed all of the ticket sale money into winnings for each game, then, at least in a cumulative sense, the purchase of lottery tickets would constitute fair bets—the average payoff would equal the average ticket price. An example of this would be if each player paid a dollar for a ticket that went into a hat, and then a winning ticket was chosen from the hat, with the purchaser of that ticket winning all of the money that had been collected. The reality is usually more complicated. Typically, multiple players can purchase the same ticket (thus having to share the winnings if that ticket is drawn) or the winning ticket might not have been purchased by anyone. In the latter case, the money is rolled over to the next game, which might be better than fair for the players if the jackpot is larger than the total investments for that week. Usually, however, the game is worse than fair for the players, primarily because the state (or whatever organization is hosting the lottery) keeps a portion of the proceeds. The state of Wisconsin, for example, pays out slightly more than half of its lottery revenue as winnings; most of the remaining revenue is used for property tax relief. Other common uses for funds among state-run lot-

History of Lotteries

In Athens during the fourth and fifth centuries B.C.E., lotteries were used to select political office holders. In Rome, the emperor Gaius Julius Caesar Augustus rebuilt his empire's infrastructure with money raised through lotteries. Lotteries also helped to fund the building of the Great Wall of China. Governments throughout much of Europe, notably in England and France, have raised essential funds with lotteries over the past few centuries.

George Washington supported lotteries as a means of funding transportation and educational systems in a fledgling nation. In the United States in the twenty-first century, most states sponsor lotteries. The jackpots for Powerball and for Mega Millions, two popular multistate lotteries, sometimes run into the hundreds of millions of dollars.

(Photos.com)

teries include education, transportation, construction, and, ironically, help for compulsive gamblers.

Calculating the Chances

Regardless of the question of fairness, a lottery is clearly disadvantageous to almost every player. Nevertheless, lotteries attract large numbers of players because people are willing to pay a small amount of money for the small chance of winning a fortune. Powerball, operated by the Multi-State Lottery Association, provides a good illustration. There are nine ways to win with a $1 Powerball ticket; in four of these ways, the winnings are less than $10. The probability of winning something is about 1:35, but the probability of winning anything more than $100 is less than 1:700,000. The probability of winning the big jackpot is 1:195,249,054, as can be verified with some basic rules of counting.

Each Powerball ticket consists of five distinct numbers, 1–59, together with a "Powerball" number, 1–39. To determine the winning ticket, five balls are randomly drawn from a drum containing white balls numbered 1–59, and then one ball (the Powerball) is drawn from a drum containing red balls numbered 1–39. The winning ticket must match all five white balls (irrespective of the order in which they are drawn) as well as the red ball. The probability of winning the jackpot is 1 divided by the number of distinct possible tickets (the number of possible outcomes of the drawing). There are 59 possibilities for the first white ball; for each of those there are 58 possibilities for the second white ball. Continuing, there are 57 possibilities for the third, 56 for the fourth, and 55 for the fifth. If the order of drawing these balls were relevant, a total of $59 \times 58 \times 57 \times 56 \times 55 = 600{,}766{,}320$ ways of drawing the white balls would be counted. This number, however, is much larger than the true probability, since the order of the drawings is not relevant. For instance, the possible outcome 2, 4, 8, 16, 32 should be counted once; but among the aforementioned count of 600,766,320, this collection of balls appears $5 \times 4 \times 3 \times 2 \times 1 = 120$ times (because ball 2 could be listed in any one of five positions, and then ball 4 could be listed in any of the remaining four positions, and so on). The earlier count should be divided by 120 in order to correct for this systematic overcounting. Finally, incorporating the possibilities for the red ball, the result should be multiplied by 39. This calculation yields the 195,249,054 possible jackpot tickets.

Winning Strategies?

One way to improve the chances of winning is to buy more tickets. A properly run lottery does not lend itself to winning strategies. For instance, the Powerball drawings are videotaped and audited, and the equipment is stored in a vault and meticulously tested for nonrandom behavior. So bribery would be difficult, and knowledge of historical winning numbers would most likely be pointless. One could ensure a win by purchasing all possible tickets (an attractive option if the jackpot has grown very large because of rollovers), but this would require a huge initial investment, and it would be quite difficult from a practical standpoint to orchestrate the purchase. Further, if multiple people purchased the winning ticket, then the jackpot would be divided among them. Commonly chosen tickets involve previous winning combinations, numbers below 32 (because they could represent birthdays or other significant dates), and simple combinations such as 1, 2, 3, 4, 5, 6. The one bit of control a lottery player does have is to avoid such combinations to reduce the likelihood of splitting the jackpot in the event of a win.

Further Reading

Bialik, Carl. "Odds Are, Stunning Coincidences Can Be Expected." *Wall Street Journal* (September 24, 2009). http://online.wsj.com/article/SB125366023562432131.html.

Hicks, Gary. *Fate's Bookie: How the Lottery Shaped the World*. Stroud, England: The History Press, 2009.

North American Association of State and Provincial Lotteries (NASPL). "Cumulative Lottery Contributions to Beneficiaries." http://www.naspl.org/index.cfm?fuseaction=content&PageID=74&PageCategory=74.

John Beam

Magic

Category: Arts, Music, and Entertainment.
Fields of Study: Algebra; Number and Operations; Representations.
Summary: Many tools of mathematics and mathematical properties lend themselves to tricks.

Mathematical magic may seem to be either redundant or an oxymoron. Many people equate mathematical processes or theorems with magic, such as the magic of logarithms or when mathematicians are thought to have magical powers with numbers floating around their heads in movies and on television. Others view it as a collection of sterile algorithms absent of any signs of magic. However, the realm of mathematical magic counters both of these views, blending together elements from mathematics as a structure with an element of surprise akin to magic. Invoking mathematics of great breadth—arithmetic, number theory, algebra, geometry, and topology—the mathematical magician's "tools" are numbers, cards, string, dice, dominoes, calendars, watches, coins, dollar bills, and rubber bands.

Arithmetic Magic

Arithmetic magic depends on the clever use of divisors, multiples, and basic operations. As an example, ask a friend to write down his or her age. Then, add the age on the friend's next birthday. Add 9 to this sum. Divide that sum by 2. Finally, subtract the friend's current age. Then, magically announce that the answer is 5. It will always be 5, thanks to mathematics. For example, if the friend's age is 24, the friend would calculate: $24 + 25 = 49; 49 + 9 = 58; 58 \div 2 = 29; 29 - 24 = 5$. In fact, with a slight modification of the first calculation (add one more than your starting number), your friend could start with any number, such as 3.5, π, or even −72.3, and the result will still be 5.

Card and Dice Magic

Mathematical magic using playing cards capitalizes on their properties—numerical values 1–13, four suits, two colors, front-back orientation—as well as the fact that a deck of cards can be both ordered and shuffled. As another example with a friend, shuffle a deck of cards, hand it to your friend, and then casually write something on a piece of paper, which is folded and set aside. Ask your friend to deal the top 12 cards face-down on the table and then touch any four cards, which you turn over. Group the other eight dealt cards and return them to the bottom of the card deck. Suppose the four face-up cards are a 3, 5, 7, and King (where all face cards are to be treated as a 10). Taking the deck, deal more cards on top of each card to make 10, counting out loud the sequences (for example, 3, 4, 5, 6, 7, 8, 9, 10 and 5, 6, 7, 8, 9, 10 and 7, 8, 9, 10). Because the King has the value 10, no cards are dealt on top of it. Hand the deck to your friend, ask him to add the values of the original four cards ($3 + 5 + 7 + 10 = 25$) and then count out that number of cards (25 cards). When the last card is turned over, reveal that it matches your prediction written on the paper.

Mathematical magic using dice depends on the fact that the pips on the opposite sides sum to 7. As an example of a trick, with your back turned, ask a friend to throw three dice on a table and add the top faces (for example, $2 + 4 + 5 = 11$). Then ask the friend to pick up any one of the dice and add its bottom number to the current sum (for example, opposite the 2 on the first dice is a 5, so $11 + 5 = 16$). Finally, ask the friend to roll that die again, and add the new top face to the current sum (for example, $16 + 6 = 22$). Turn around and announce that you have no way of knowing which die was rolled twice, pick up the 3 dice, shake them in your hand, and magically announce your friend's final sum.

Geometric Magic

Mathematical magic involving geometry or topology is similar to actual tricks performed by magicians, such as the Chinese Linking Rings, Magical Knots, and Houdini Escapes. As a simple example, start with an 8 × 8 grid square and draw 3 lines to subdivide it as shown. Cut along the 3 lines, producing 4 pieces, which can be rearranged to form the 5 × 13 solid rectangle. What is the magic? The initial square with an area of 64 square units has been transformed into a rectangle with an area of 65 square units.

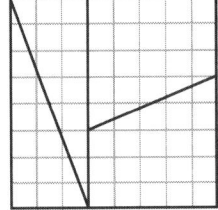 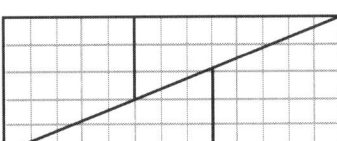

The Magic Revealed

Why do the previous four tricks work? The first arithmetic trick is explained using algebra, where N is the starting number, shown as

$$\frac{N + (N + 1) + 9}{2} - N = 5.$$

For the second trick, it is important that the card you write on the paper matches the bottom card on the shuffled deck at the start. The trick becomes automatic, since the 4 face-up cards and the 8 cards placed on the bottom as part of the deck essentially force your "secret card" to now be in the 40th position in the original deck. The counting mechanism forces this card to be the card revealed. For the third trick, determine the final sum by adding 7 to the sum of the 3 top faces seen as you pick up the dice. Finally, for the fourth trick, the magical effect is because of the apparent diagonal of the rectangle, as it is not a straight line but is a "thin" parallelogram with an area of 1 square unit. To show this mathematically, the two line segments forming the diagonal have differing slopes of 3/8 and 2/5. As a twist to this trick, note that the square had side length 8 while the rectangle had side lengths 5 and 13, where the numbers 5, 8, 13 are part of the Fibonacci sequence. In fact, any three ordered numbers (different) in this sequence produces this magical effect.

Magic Squares, Cubes, and Circles

In any discussion of mathematical magic, one must mention magic squares, cubes, and circles. First, subdivide a square into smaller squares, each containing a number. The magical effect is that the numbers in each row, each column, and each diagonal all sum to the same constant value.

8	1	6
3	5	7
4	9	2

This common example is the "Lo Shu" magic square with a constant sum of 15, being part of the legend (650 B.C.E.) of the Chinese Emperor Yu finding a turtle with the same square inscribed on its back. Also, the German artist Albrecht Dürer inserted a famous magic square in his painting *Melancholia*, with its constant sum of 34 and the painting's date of 1514 included in the bottom row of cells.

16	3	2	13
5	10	11	8
9	6	7	12
4	15	14	1

Historically, mathematics and magic are intertwined, back to the Pythagoreans who revered certain numbers with a special mysticism. This "aura" of numbers having special magical effects surfaced often throughout history in the form of special primes, special products, and special properties. For example, one can not dismiss the magic of numbers when considering these number patterns, all evoking a feeling of "Behold!"

$$0 \times 9 + 1 = 1$$
$$1 \times 9 + 2 = 11$$
$$12 \times 9 + 3 = 111$$
$$123 \times 9 + 4 = 1111$$
$$1234 \times 9 + 5 = 11111$$
$$12345 \times 9 + 6 = 111111$$
$$123456 \times 9 + 7 = 1111111$$
$$1234567 \times 9 + 8 = 11111111$$
$$12345678 \times 9 + 9 = 111111111$$
$$123456789 \times 9 + 10 = 1111111111$$

$$1 \times 8 + 1 = 9$$
$$12 \times 8 + 2 = 98$$
$$123 \times 8 + 3 = 987$$
$$1234 \times 8 + 4 = 9876$$
$$12345 \times 8 + 5 = 98765$$
$$123456 \times 8 + 6 = 987654$$
$$1234567 \times 8 + 7 = 9876543$$
$$12345678 \times 8 + 8 = 98765432$$
$$123456789 \times 8 + 8 = 987654321$$

$$9 \times 9 + 7 = 88$$
$$98 \times 9 + 6 = 888$$
$$987 \times 9 + 5 = 8888$$
$$9876 \times 9 + 4 = 88888$$
$$98765 \times 9 + 3 = 888888$$
$$987654 \times 9 + 2 = 8888888$$
$$9876543 \times 9 + 1 = 88888888$$
$$98765432 \times 9 + 0 = 888888888$$

$$12345679 \times 9 = 111111111$$
$$12345679 \times 18 = 222222222$$
$$12345679 \times 27 = 333333333$$
$$12345679 \times 36 = 444444444$$
$$12345679 \times 45 = 555555555$$
$$12345679 \times 54 = 666666666$$
$$12345679 \times 63 = 777777777$$
$$12345679 \times 72 = 888888888$$
$$12345679 \times 81 = 999999999$$

$$1^2 = 1$$
$$11^2 = 121$$
$$111^2 = 12321$$
$$1111^2 = 1234321$$
$$11111^2 = 123454321$$
$$111111^2 = 12345654321$$
$$1111111^2 = 1234567654321$$
$$11111111^2 = 123456787654321$$
$$111111111^2 = 12345678987654321$$

Martin Gardner claimed in his 1956 book *Mathematics, Magic and Mystery* that mathematical magic has a unique but limited audience. In his opinion, mathematicians reject mathematical magic as trivial and dull, while magicians reject it as pseudomagic. The true audience is therefore those who appreciate mathematical recreations implemented in a creative, entertaining context. A master of such presentations is Arthur Benjamin, a combinatorics professor and professional magician, who has appeared on many radio and television programs, such as the widely popular political satire program *The Colbert Report*, and been profiled in entertainment, news, and scientific publications. His popular demonstrations and explanations of methods for rapid mental calculations, which have been enjoyed by audiences of all ages and cultures worldwide, as well as his many popular books on mathematical magic would appear to belie Gardner's claim.

Further Reading
Andrews, W. S. *Magic Squares and Cubes*. New York: Dover, 1960.
Benjamin, A. *Secrets of Mental Math: The Mathemagician's Guide to Lightning Calculation and Amazing Math Tricks*. New York: Three Rivers Press, 2006.
Blum, Raymond. *Mathemagic*. New York: Sterling, 1992.
Carter, Philip, and Ken Russell. *The Complete Book of Fun Maths: 250 Confidence-Boosting Tricks, Tests and Puzzles*. Mankato, MN: Capstone, 2004.
Gardner, Martin. *Mathematics, Magic and Mystery*. New York: Dover, 1956.
Longe, Bob. *The Magical Math Book*. New York: Sterling Publishing, 1997.

Jerry Johnson

Marriage

Category: Friendship, Romance, and Religion.
Fields of Study: Algebra; Data Analysis and Probability.
Summary: Sociologists and others have made many demographic studies of marriage, even modeling it.

Many kinds of arrangements have existed throughout history under the umbrella of marriage, with the expectations and responsibilities of married partners and their rights both to enter into marriage and within the marriage changing considerably over time and across (or within) cultures. It has always included legal and economic dimensions, which have played into the changing demographics of the married.

History of Marriage

The modern concept of "marriage for love" is a relatively recent development in the history of marriage; for several millennia, marriage was an important societal convention fulfilling critical economic, legal, and political functions. Among elite people, marriage was a tool for the control and consolidation of wealth and power by forming strategic alliances between families. Political and military agreements were sometimes forged in the context of a marriage. In middle and lower classes, marriage played a similarly important societal role, especially economically. Marriage's economic role was further reflected in conventions such as illegitimacy, the dowry, and large families of children, which proved a vital source of labor and economic gain for the family. Marriage was also the societal device for conferring a host of legal rights.

The sexual marriage, a marriage that is freely arranged between two people on the basis of love, is a newer development that evolved from cultural changes that occurred during the Enlightenment and were further developed by the Industrial Revolution. The economic and legal changes that grew from this period gradually eroded the historical reasons behind arranged marriages. This gradual change in marriage perhaps culminated with the 1950s concept of the "Leave It To Beaver family"; however, this short-lived paradigm of marriage experienced dramatic shifts in the socially turbulent decades to come.

The legal and political advances for women in the early twentieth century, coupled with important economic and demographic advances in the latter half of that century, paved the way for important changes in the way people approach marriage. Women made significant strides economically and socially that allowed them the possibility of viable, independent lives apart from marriage. The innovation of birth control also played an important role in the evolution of marriage by allowing women to effectively separate sex and child rearing.

Statistically Analyzing Marriage

Marriage in the United States has undergone critical demographic changes that are closely allied with education level and socioeconomic status. Data spanning five decades of the latter twentieth century and early twenty-first century demonstrate a steady decline in marriage rates. In 1970, 84% of adults aged 30–44 years were married compared with only 60% in 2007. The decline in marriage rates mirrors corresponding rises in the divorce rate and a greater tendency of couples to find alternate arrangements, such as short-term relationships and cohabitation.

Marriage is associated with well-established economic benefits. Most obvious is the economy of scale realized when a couple can share major assets, like a house, a car, or furniture, that they would otherwise each need to purchase individually. This economy of scale is still a significant advantage even when the additional economic cost of raising children is factored in.

Table 1. Percentage of married adults 30–44.

1970	84%
1980	77%
1990	69%
2000	65%
2007	60%

However, economic benefits are also realized when a spouse marries someone with a higher income. In 2007, individual income for married men was an average of 12% higher than for single men. Married women outearned their single counterparts even more substantially, with a 53% higher average income. However, this statistic is not a simple causal relationship between being married and accruing greater wealth; these economic gains are closely tied to education level and earning power. Essentially, people with a higher educational level are more likely to be married, more likely to be married to a spouse of a similar educational level, and more likely to realize and compound the economic benefits of marriage. Interestingly, this is a trend not present in the 1970 data, where the marriages rates across the socioeconomic spectrum were nearly identical. The period since the 1970s has seen significant changes in the number of women attending college and their choices in forming relationships.

Research literature also indicates important health and emotional benefits associated with marriage. These benefits stem not only from lifestyle changes (for example, the healthier diet of a married couple or the shared division of household labor); contemporary studies suggest an even more important factor is the mitigation of stress and its effects on health. Married people live longer, experience less illness, and are less prone to many diseases. Importantly, studies clearly indicate that the quality of the marriage is an important factor; people in poor marriages have been shown to be even unhealthier than single people. There are also clear gender differences in the extent and the way in which spouses realize the health benefits of marriage.

Mathematically Modeling Marriage

Marriage statistics are extensively tabulated like many other social statistics, but researchers also use mathematical modeling to study marriage. The 2003 book *The Mathematics of Marriage: Dynamic Nonlinear Models* was authored by an interdisciplinary team including mathematicians. It used mathematical ideas, such as difference equations, phase space, null clines, influence functions, inertia, and stable steady states (attractors), to model marriage, with applications to other psychological phenomena. In 2009, a team of mathematicians from the United Kingdom and the United States analyzed the behaviors of 700 couples over the course of 12 years to develop a probabilistic model that accurately predicted which marriages would last. It was based on classifying couples into one of five types using behavioral variables. Only one type suggested a long-lasting marriage. In 2010, Spanish economist José-Manuel Rey developed an equation based on optimal control models and the "second thermodynamic law for sentimental interaction," which states a relationship will disintegrate unless it receives input "energy" or effort.

As with the Birthday and Cocktail Party Problems, mathematicians have identified a similar social puzzle in the Stable Marriage Problem. First introduced as a matching problem by D. Gale and L. S. Shapley in 1962, the stable marriage problem consists of equal numbers of single men and women. Every man creates a preference ranking of each woman as a potential match; similarly, every woman ranks each of the men. The goal is to pair the men and women in couples so as to create stable, happy marriages.

The technical challenge is to avoid an "unstable matching," which arises when a man and woman who are not paired under the matching would each prefer to be with each other over their paired spouse. The immediately interesting question—whether there always exists a stable matching given a set of preference rankings for each individual—was answered in the same seminal work. The Stable Matching Algorithm provides a solution to this problem and, furthermore, is guaranteed to always produce a stable matching. Curiously, this algorithm maximizes one gender's happiness while minimizing the other's, depending upon which gender does the proposing and which does the accepting. This same algorithm has other applications, for example, in matching medical school applicants with schools and in pairing roommates for college residence halls.

Further Reading

Coontz, Stephanie. *Marriage: A History*. New York: Penguin. 2005.

Fry, Richard and D'Vera Cohn. "Women, Men and the New Economics of Marriage." Pew Research Center (January 19, 2010). http://pewsocialtrends.org/010/01/19/women-men-and-the-new-economics-of-marriage.

Gale, D., and L. S. Shapley. "College Admissions and the Stability of Marriage." *American Mathematical Monthly* 69 (1962).

Gottman, John, James Murray, Catherine Swanson, Rebecca Tyson, and Kristin Swanson. *The Mathematics of Marriage: Dynamic Nonlinear Models*. Cambridge, MA: MIT Press, 2003.

Knuth, Donald. *Stable Marriage and Its Relation to Other Combinatorial Problems: An Introduction to the Mathematical Analysis of Algorithms*. Providence, RI: American Mathematical Society. 1997.

Parker-Pope, Tara. *For Better: The Science of a Good Marriage*. New York: Dutton, 2010.

Matt Kretchmar

Martial Arts

Category: Games, Sport, and Recreation.
Fields of Study: Algebra; Geometry.
Summary: The motions and stances of martial artists can be analyzed for their efficiency and use of force.

In the martial arts, humans use repetitive training to standardize their response to threat. The original bare-handed style of ritualized combat training that evolved into the modern martial arts is believed to have developed in China at about the same time as the introduction of bronze, agricultural sciences, and Chinese philosophy, and later spread to Korea and Japan. Many regions of the world have their own native forms of combat training, which are now also called "martial arts" in English, but the English term comes originally from the Japanese.

While techniques and philosophies differ, the underlying goal of all martial arts is the same: that through deliberate physical and mental training, forces can be concentrated or dissipated across time and space in order to either attack or defend. In the modern world, most martial artists train for sport or health promotion. Mathematics can be used to describe and model the stances and movements of martial arts forms and practitioners, such as the geometry of balance and the forces concentrated across time and space in the form of kicks, blocks, and strikes.

Etymology

The term "martial arts" first appeared in English in 1933. The Japanese Railway Ministry released the *Official Guide to Japan*, including a reference to the Butokukai in Kyoto, which they translated as the "Association for Preserving the Martial Arts." "Martial arts" became an umbrella term describing the fighting skills displayed by Japanese practitioners of jiu jitsu who had been invited to give demonstrations in England and the United States in the late nineteenth century and the judo practitioners who followed soon after. When American troops returned from the occupation of Japan in the 1940s and 1950s, they brought along some knowledge of and interest in karate. In the 1960s, the Chinese Martial Arts came to be recognized in the West and were grouped under that increasingly pan-Asian umbrella term. Since then, many modern and traditional martial arts have been recognized to varying degrees, and the term has become international.

The Mathematics of Attack and Defense

An attack is the concentration of force across time and space. An ideal blow multiplies the mass of the entire body by the speed at which the striker moves and delivers the resultant force to a precisely determined surface. This may be done to inflict damage directly or to interfere with the opponent's intent by disrupting his or her balance. Defense is the opposite, dissipating the attacking force across both time and space by either absorption, deflection, preemption, or avoidance. The same principles that allow defense against an attack can be used to dissipate an entire conflict.

Variables that affect the force delivered or deflected include center of gravity or mass, kinetic energy, linear and angular (rotational) momentum, velocity, inertia, and acceleration (as governed by Isaac Newton's laws of motion). Mathematicians have studied and modeled many aspects of martial arts. Analysis of data has shown that kicks are typically three to six times as powerful as punches; the speed of a fist during a forward punch is a nonlinear function of arm extension; and a smaller fighter can punch as hard as a larger one by moving faster. Some of these models approximate body parts with geometric forms, such as cylinders for arms, in order to simplify the calculations involved. Geometry is also important for examining the basic stances and movements of all martial arts. Stability for both attack and defense comes from maintaining the correct alignment and balance in three dimensions. The mathematics becomes even more complicated once the practitioner starts moving. Correct form requires a specific angle between body parts when kicking or punching. These angles have been determined through generations of practice and can be measured very specifically by the avid student who enjoys applied mathematics. In this way, experts in many martial arts have learned that correcting the angle of one's foot or knee or wrist by just a few degrees makes all the difference for gaining leverage or applying the maximum amount of force. These small differences, best measured mathematically, can make the difference between a novice and a martial arts master.

In martial arts, a kick is generally three to six times as powerful as a punch. (Photos.com)

There are many martial arts, but they all present both attacker and defender with the challenge of maintaining one's own intent while interfering with the intent of one's opponent. This is like balancing an equation, where the intent of the two or more people involved in a confrontation can be reduced like the terms in an exercise in algebra. A parry on one side negates a strike from the other, and so on. This is why martial artists are sometimes seen standing almost still and looking at each other before a fight begins. In their minds, they are balancing out the equation. Usually this ends when one or the other thinks he or she see a way to make the balance work out in their favor and they start the action. Sometimes, however, the equation is so unbalanced that both sides can see it, and the fight ends without any violence at all.

Further Reading

Diacu, Florin. "On the Dynamics of Karate." http://www.math.ualberta.ca/pi/issue6/page09-11.pdf.

Sprague, Martina. *Fighting Science: The Laws of Physics for Martial Artists*. Wethersfield, CT: Turtle Press, 2002.

Starr, Phillip. *Martial Mechanics: Maximum Results With Minimum Effort in the Practice of the Martial Arts*. Berkeley, CA: Blue Snake Books, 2008.

John N. A. Brown

Mathematical Puzzles

Category: Games, Sport, and Recreation.
Fields of Study: Algebra; Geometry; Reasoning and Proof.
Summary: The emphasis on problem solving in mathematics lends itself well to puzzles.

When considering mathematical puzzles, there are really two different types of puzzles available. Some puzzles are mathematical in nature, but require no mathematics to solve—similar to games like checkers, chess, and tic-tac-toe. Other puzzles are mathematical in nature and require a certain level of mathematics to solve—similar to games like cryptograms and Sudoku. Sometimes mathematical puzzles are referred to as "brainteasers."

Tower of Hanoi

One of the oldest mathematical puzzles is the Tower of Hanoi. This puzzle was developed in 1883 by French mathematician Édouard Lucas. In the game, the player has several disks of different sizes and three pegs. The object is to move all of the disks from the starting peg to a different peg, according to the rule that a disc can only be placed on an empty peg or on top of a larger disc. In the legend believed to have inspired the game, there is a Vietnamese temple in Hanoi that contains a large room with three posts surrounded by 64 golden disks. The temple priests perpetually move the disks, according to the rules of the puzzle. According to the legend, when they are done, the world will end. If the legend were true, and if the priests moved disks at a rate of one per second, it would take them a minimum of 18,446,744,073,709,551,615 turns to finish—585 billion years. In general, the number of starting disks will determine the minimum number of moves to solve the puzzle.

To move a single disk requires only one move. To move two disks (D_1 and D_2 with the smaller number being the smaller, or topmost, disk) would require three moves: (1) D_1 to an empty, (2) D_2 to an empty, and (3) D_1 onto D_2. Three disks would require seven moves: move the top two disks as described above (three moves), move the last (bottom) disk to the empty, then move the two-disk stack onto the third disk (another three moves). A fourth disk would similarly require $7 + 1 + 7 = 15$ moves. Using this pattern, the minimum number of moves for an additional disk will be double what the previous number of layers took plus one. However, to find the minimum number of moves for 10 disks, one needs to know what the minimum number of moves for nine disks would be. For nine disks, one needs to know the minimum number of moves for eight disks, and so on. Although a working recursive formula exists, it is not helpful for large numbers of disks. However, there is a pattern that can be found looking at the minimum number of moves for a certain number of disks that can be used to determine the minimum number of moves for any number of disks. In general, if there are n disks, the minimum number of moves to solve the tower problem will be $2^n - 1$.

Two-Container Problem

Another old mathematics puzzle that was used in the 1995 movie *Die Hard with a Vengeance* involves two containers of different sizes that are used to measure a different third value. For example, in the movie, the characters were given a 5-gallon and a 3-gallon container and needed to measure exactly 4 gallons of water. It is assumed that there is an unlimited amount of water to pour into either container, and that contents of either container can be poured down a drain. Other versions of this puzzle can be formed by changing the size of the original containers or the quantity needed at the end. If the containers have capacities that are relatively prime to one another (greatest common factor is one), then any number less than the bigger container can be achieved. If the capacities are not relatively prime, then only certain values can be obtained. For this specific version, if x equals the number of times the 5-gallon container is filled and y equals the number of times the 3-gallon container is filled, the problem can be rewritten as an equation in two variables: $5x + 3y = 4$.

Any ordered-pair solution to this equation will be a solution to the problem, although the method would still have to be determined. In the movie, the solution they found was $(2, -2)$. The five-gallon bottle needed to be filled two times and the three-gallon bottle needed to be emptied twice (hence, the negative number). To actually solve the problem, they would have to fill the five-gallon container (first fill) and use it to fill the three-gallon container, leaving two gallons in the five-gallon container. The three-gallon container would then be emptied (first empty) and the remaining two gallons poured into the three-gallon container. The five-gallon container would then be filled again (second fill) and used to pour into the three-gallon container. Since the three-gallon container would have two gallons of water already inside, it would only hold one more gallon, leaving four gallons in the five-gallon container. The three-gallon container would then be emptied (second empty), leaving exactly four gallons. An alternate solution to this equation is (–1, 3).

Cabbage, Goat, Wolf

Another type of mathematical puzzle involves three objects and a keeper. As long as the keeper is present, all objects will remain safe, but if the keeper were to leave certain pairs of objects together unsupervised, at least one would be destroyed. For example, a farmer needs to transport cabbage, a goat, and a wolf across a river. The farmer is the only one who can row the boat and the boat is only large enough to carry the farmer and one other object. The goat and the cabbage cannot be left alone together as the goat would eat the cabbage. Similarly, the wolf and the goat cannot be left together as the wolf would eat the goat. The wolf has no interest in the cabbage, so that pair can be left alone together. The task is to determine how the farmer will get all three objects across the river.

On the initial row, the farmer's only option is to take the goat. If he takes the cabbage, the goat is eaten. If he takes the wolf, the cabbage gets eaten. Once the goat is on the other side, the farmer leaves the goat and returns across the lake alone. The farmer must now choose to take either the cabbage or the wolf to the other side. The farmer returns to the first side with the goat and swaps the goat for the last object on the original side. Upon crossing the river, the farmer now leaves both the cabbage and the wolf on the opposite side of the river and returns to the original side with an empty boat in anticipation of picking up the goat. One final row allows the farmer and all three objects to be on the far side of the river.

Squaring a Double-Digit Number

Some mathematics puzzles take the form of mathematics magic. For example, if a spectator calls out any two-digit number, the mathematician can square the number without a calculator in a short amount of time—with practice, faster than a human verifying it on a calculator. Finding the square of some numbers is easy; for example, any multiple of 10 (such as 10, 20, or 30). All that is needed is to square the 10s digit and concatenate two zeros to the right. For instance, 70 squared would be 4900. A number that has a five in the ones digit is also easy to square; merely take the 10 digit, multiply it by the next-highest integer, and concatenate a 25 to the right. For example, to find 75 squared, take $7 \times 8 = 56$, then append 25 to get 5625. However, there are 90 possible two-digit numbers that could be called out and only 18 that fit one of the patterns above. For the remainder, the mathematician can employ a principle referred to as "squaring a binomial," which is expressed algebraically as

$$(A+B)^2 = A^2 + 2AB + B^2.$$

If one needs to square a different two-digit number, such as 43, mentally rewrite 43 as $(40+3)$. Using the above formula, the square can be found by

$$43^2 = (40+3)^2 = 40^2 + 2(40)(3) + 3^2$$
$$= 1600 + 240 + 9 = 1849.$$

As mentioned above, 40 is a multiple of 10 and easy to square; similarly, 3 is easy to square. The more difficult part of the formula to calculate in one's head is the middle—take 40 times 3 and double it. Then, add those three numbers together to get the square of the original number.

Squaring a number that has a 5 in the ones digit is a special case of squaring the binomial. If t equals the tens digit, then $10t + 5$ is the original number. Squaring the binomial yields

$$(10t+5)^2 = (10t)^2 + 2(10t)(5) + 5^2$$
$$= 100t^2 + 100t + 25.$$

Factoring $100t$ from the first two terms yields

$$100t(t+1) + 25.$$

Martin Gardner and Recreational Mathematics

Martin Gardner (1914–2010), an American mathematician, specialized in recreational mathematical games. From 1956 to 1981 he wrote *Scientific American* magazine's Mathematical Games column and is credited by many for almost single-handedly sustaining and nurturing interest in recreational mathematics for much of the twentieth century. The kind of mathematical games

Gardner wrote about are still being promoted not only for training children's minds for mathematics, both in and out of school, but also for helping older citizens maintain sharp minds. In addition to paper and pencil books, there are many Web sites aimed at seniors that have mathematical puzzle collections, and popular handheld gaming devices (like the Nintendo DS) are now being targeting at consumers in all age groups for mathematics and memory games.

Further Reading

Behrends, Ehrhard. *Five-Minute Mathematics.* Providence, RI: American Mathematical Society, 2008.

Gardner, Martin. *Hexaflexagons, Probability Paradoxes, and the Tower of Hanoi: Martin Gardner's First Book of Mathematical Puzzles and Games.* Cambridge, England: Cambridge University Press, 2008.

———. *My Best Mathematical and Logic Puzzles.* New York: Dover, 1994.

Vennebush, G. Patrick. *Math Jokes 4 Mathy Folks.* Brandon, OR: Robert Reed Publishers, 2010.

Winkler, Peter. *Mathematical Puzzles: A Connoisseur's Collection.* Natick, MA: AK Peters, 2004.

Chad T. Lower

Mathematics and Religion

Category: Friendship, Romance, and Religion.
Fields of Study: Reasoning and Proof; Connections.
Summary: The connection between religion and mathematics is intricate, spanning cultures and centuries, with mathematics itself sometimes manifesting religion-like features.

Mathematical knowledge has been intertwined with spiritual or religious contemplation since humans began to develop numerical, spatial, and symbolic reasoning in order to understand the world and humanity's place within it. Both practical and abstract knowledge have been significant to cosmological and theological considerations. Another way that mathematics is linked to religion is by those who suggest that mathematics is a religion.

Mathematics provides tools that underpin computation, prognostication, organization, and design. Consequently, mathematical knowledge—as constituted by practical arithmetical (computational), algebraic (numerical problem solving). and geometric (spatial) knowledge—has been an essential ingredient in divination as well as in ritual constructions and practices. The influences of mathematical knowledge, broadly construed, on cosmology can be found in different times, places, and cultures. They are evident in a variety of contexts that include Pythagorean, Judaic, and Chinese number mysticism; Vedic rituals; Islamic trigonometry; and pattern drawings that some South Pacific Islanders believe are essential to entering the land of the dead.

Beyond skill-based practicality, mathematics as a way of obtaining infallible knowledge of transcendental objects engendered and strengthened spiritual considerations that became more closely aligned with doctrine. It did so to such an extent that the development of new mathematical knowledge often instigated immediate responses from religious authorities. Such symbiotic yet ever-evolving relationships between mathematical epistemology and theological contemplation are a central feature of the Christian tradition across the ages.

Implicit Practices, Divination, and Pattern Drawing

In the oldest cultures it is difficult to separate mathematical and ritual practices. Shamans and priests, from ancient Babylonia to Mesoamerica, used arithmetical and geometrical knowledge as part of their efforts to organize time and space so as to facilitate particular observances. In some cultures, the drawing of geometric patterns was integral to storytelling that conveyed origin myths as well as aspects of the afterlife. For both ancient and contemporary peoples, mathematics is not identifiable as a constituent of an explicitly distinctive knowledge. Rather, mathematics as it is recognized today is seen as implicitly embedded within customs of cultural significance that included spiritual well-being.

Divination, as practiced in various times and places, typically involves both randomness and structure. The objects required for the foretelling of events, while specific to custom, are subjected to a process that produces a random outcome. The diviner's skill comes into play when interpreting the result. Doing so involves adhering to rules that apply to the particular procedure. Consequently, divination often involves strictures that can be resolved into numerical or logical systems, systems that often reflect binary considerations. Such can be found today in the methods of divination practiced by the Caroline Islanders of the South Pacific (knot divination), the Yoruba people of Africa (*Ifa*), and the Malagasy (*Sikidy*).

Pattern drawing has often accompanied cultural narratives regarding both ancestors and the afterlife. Such traditions continue into the modern era with the Tshokwe people of Angola and the Malekula of Vanuatu. In each case, intricate patterns are drawn in a continuous, uninterrupted fashion. While modern mathematics conceives of such in terms of graph theory and Eulerian circuits, there is little evidence to suggest that the cultures discussed here have an explicit or external framework within which such patterns are considered. Indeed, Tshokwe have relatively few patterns that accompany their origin myths, and knowledge of their production is limited. The cultural situation for the Malekula is considerably different. Their patterns number in the hundreds and all require that the tracing begin and end at the same point without repetition of any edge. Knowing how such patterns are produced, which constitutes a form of implicit mathematical training insofar as it recognizes various systematic elements within the drawings, is part of what men pass on to their sons. The ability to reconstruct a pattern correctly earns one access to the land of the dead.

Like the Tshokwe and Malekula, Tamil women in southern India draw patterns as a way of marking passages or transitions. The ritual designs produced by them, which are known as *kolam*, are used to decorate the entrance to a house. They vary according to the events being marked, many of which relate to life- or worship-cycles. Recently *kolam* have attracted the attention of computer scientists who are interested in the formulation and formalization of picture languages.

Classical and Judeo-Christian Traditions

The mathematics of Greek antiquity marked a distinctive break with the implicitly integrated practices associated with various cultures across time. Moreover, it laid the foundation for more explicitly considered connections between the mathematical and the spiritual.

The perspective held during the earliest portion of the Pythagorean-Platonic period is often simply characterized as follows: number is religion and religion is number. That is, numbers provided the lens through which the Pythagoreans viewed the cosmos. In this way mathematics links the mundane and the sacred in ways exemplified in different ages and cultures. Mathematical reasoning, which at this time might more closely have aligned with numerology or number mysticism, provided a means of bringing order and harmony to the universe.

The realization that certain numbers are irrational (that is, that some measures are incommensurable) represented a serious challenge to Pythagoreans, for it contradicted the assumption of a cosmic harmony. While the need to resolve paradoxes instigated monumental discoveries, it soon became apparent, as it would again in the years to come, that using mathematics as a means of demystifying the world could engender new and even greater mysteries.

According to Plato, arithmetic and geometry constituted areas of study essential to higher education, and thus they became part of the quadrivium of Western education, which included astronomy and music. That philosophical discussions found in his dialogues turn to and on mathematical reasoning underscores the significance of mathematics to Platonic conceptions of the good and true. It represented an "a priori," if for many a latent, body of knowledge through which one accessed eternal and perfect forms rather than transient and imperfect perceptions of these.

While he maintained a distinction between the physical and the otherworldly, Aristotle differed from those who believed that mathematics provided a special conduit to transcendental realms. Rather, his perspective of mathematics as abstraction based on physical reality reverses the mystical point of view. Aristotelian thinking underpins a more humanistic and, in later ages, secular understanding of mathematics. Underscoring the difference between process and object, classical Greek mathematics attempted to distinguish between the potential and the actual when discussing infinity. Powerful analytic arguments and famous paradoxes hinged on the process of infinite subdivision that gave rise to infinitesimal considerations. Amid this conceptual ambiguity, Aristotle maintained that the actual infinite—the infinite as a completed object—is unknowable.

Euclid's *Elements* is especially significant among classical texts that helped to solidify, as well as perpetuate, connections between mathematical and metaphysical reasoning. As a compendium of geometric knowledge of its day, *Elements* is most significant for its presentation of timeless and unassailable conclusions rigorously deduced from self-evident truths. It speaks to absolute certainty and provides geometry as a model for attaining such. Consequently, the influences of the *Elements* on mathematics and Christian theology echo across the centuries.

Aurelius Augustinius (354–430), or Saint Augustine, helped to begin the process of transforming Pythagorean–Platonic conceptions into Christian doctrines during the Middle Ages (fifth through twelfth centuries). His contributions, among many things, served to imbue Christian symbolism, including the Ark of the Covenant with its divinely prescribed dimensions, with numerical and geometric significance. Such symbolism was considered necessary for analogizing and simulating the majesty of God's power in ways comprehensible to a faithful laity. Following classical traditions, numbers represented an ideal conduit for transcendental contemplation. Shapes, on the other hand, could both signify the sacred and convey divine wisdom. The successful adaptation of Hellenistic mathematical cosmology to Christian theology owes much to Saint Augustine and others.

Scholastic theologians of the Early Modern period (twelfth through sixteenth centuries) built upon the connections between mathematics and Christian faith promoted by Saint Augustine. Setting the tone for the age, Giovanni di Fidanza (1221–1274), or Saint Bonaventure, extended Aristotle's prohibition against attempting to understand the infinite by claiming that it existed in God only. Even so, one could aspire to a better appreciation of the divine. To this end Nicholas Cusanus (1401–1464), or de Cusa, believed that mathematics emulates the creative power of God insofar as it is a manifestation of humankind's ability to create knowledge and to completely understand this creation. By virtue of this manifestation, mathematics served as an essential and mutually beneficial component of Cusanus's theology. Specifically, practicing mathematics is a way by which humankind can become closer to the divine. Whereas the platonic dialogues use mathematics to underpin conceptions of the Good, the Neoplatonic theology of Cusanus redirects mathematical attention toward conceptions of the divine.

Rendering perspective in painting by means of a vanishing point is one of the most important markers of Renaissance art. Anticipating the aesthetic significance of this development, Roger Bacon (1214–1294) encouraged the incorporation of geometric innovation in painting, believing it offered a way of better communicating God's majesty through more powerful visual imagery. As such sentiments make clear, the connections between mathematics and religion could be both rendered and read visually, thereby making such concepts accessible to lay audiences who were not necessarily conversant with the particulars of either.

Alongside Neoplatonic scholasticism, the late Middle Ages saw a resurgence of interest in gematria, a practice by which one attempts to reveal and interpret divine secrets through the association of alphabetic characters with numbers. Truth seeking by means of numerically organized systems was not a new development; it has a long history in the Jewish religious tradition and is central to Kabbalism. Among the more shocking identifications established by Michael Stifel (1486–1567) through gematria was Pope Leo X with the Beast of the Apocalypse. Similar ideas underpin recent interest in topics such as the Bible Code.

While breaking with the intellectual traditions of the past, mathematicians associated with the Scientific Revolution (c. sixteenth through eighteenth centuries) and the Modern period (from c. eighteenth century) continued to connect the discipline's reasoning and knowledge with theological concerns. René Descartes (1596–1650) promoted the individual's power of reason through geometry. His rationalism was a reaction against the constraints of scholasticism and, therefore, many considered it a threat to religious authority. Nevertheless, and like others of the age, including Gottfried Leibniz (1646–1716), he used humankind's ability to reason mathematically as the basis for discussions that ultimately asserted the existence of God.

Unlike some inclined to rationalism and deism, Blaise Pascal (1623–1662) believed that mathematical reasoning could not be applied to prove the existence of God. Another critic of mathematics' influence on theology, George Berkeley (1685–1753) pointed out that accepting the mysterious notion of infinitesimal quantities so essential to the development of calculus was tantamount to an act of faith. Consequently, he contested deism by asserting that mathematical knowledge could not provide a more exact, or more acceptable, model for theological reasoning.

Immanuel Kant (1724–1804) asserted that geometry is a contentful, or synthetic, knowledge that adheres to a universal, "a priori" form of spatial intuition. He did not, however, use this to gird theological speculation. Indeed, he attacked proofs of God's existence in his *Critique of Pure Reason* (1781) and *Critique of Practical Reason* (1788). Rather, Kant posited morality as a distinct form of intuition. The knowledge built upon this intuition leads to an understanding of the divine. Though independent forms of intuition, the geometric and the moral knowledge built upon these exemplified a common epistemological perspective.

The power of Kant's argument is evident in responses to the development of non-Euclidean geometries in the nineteenth century. With this development, the absolute certainty long associated with geometric reasoning gave way to contingent knowledge. Along with more familiar paradigm shifts, most notably Darwinian evolution, new mathematical knowledge contributed significantly to the Victorian crisis in faith. Euclid's *Elements* anchored mathematical and theological speculation for centuries; its promise of eternal and necessary truths was much in doubt.

Considerations outside geometry also exacerbated religious anxieties. Though obsessed with the notion of an all-encompassing infinite informed by the *Ein Sof* of the Jewish religious tradition, Georg Cantor (1845–1918) further destabilized relations between mathematics and spirituality with investigations that sought to establish the cardinality of the real continuum. Correspondences with Pope Leo XIII provide evidence that Cantor himself was concerned with the contentious potential of his work. The distinction between process and object so clearly delineated in antiquity meant that Christianity could safely adjudicate conceptions of the infinite as these pertained to the divine. Cantor's identification of infinite sets as objects of mathematical interest represented a clear threat to this religious privilege.

Some claim that new and contingent perceptions of mathematical certainty evident from nineteenth-century innovations instigated a period of desecularization. Failure to secure mathematics on a firm epistemological foundation through Formalism, Logicism, and Constructivism suggested that its knowledge is the confirmation of intuitions and creative possibilities,

Mathematics as Religion

Modern mathematics is seen by some people to have features similar to those of religion; for example, that mathematical foundations are accepted on belief rather than logic and comprehension. Analogies between mathematics and religion also include discussions about their omnipresent nature, their pivotal role in society, and their dependence on teaching the next generation. Some people point to aspects of mathematics that may appear unresolved or include contradictions, such as the axiom of choice or Kurt Gödel's incompleteness theorems, which showed that there are limitations to axiomatic systems. For example, in 1999 John Barrow wrote, "If a 'religion' is defined to be a system of ideas that contains unprovable statements, then Gödel has taught us that, not only is mathematics a religion, it is the only religion that can prove itself to be one."

Religious terms have been applied to mathematical theorems or mathematicians. For example, mathematical discoveries are sometimes described in terms of a revelation, epiphany, or heresy. In 1985, mathematician Paul Erdös asserted that it was important to believe in *The Book*, an imagined type of bible containing elegant proofs. This assertion inspired the 1998 work *Proofs from The Book* by Martin Aigner and Günter Ziegler. Some people have referred to mathematicians, including Erdös, as priests of mathematics who share their gospel. Mathematicians, philosophers, and theologians also consider whether a divine force is needed to explain such concepts as how the universe was formed or whether the underlying mathematical and physical principles are sufficient, which is Stephen Hawking's assertion in his 2010 book *The Grand Design*.

even if such cannot be constrained by any particular formal systems. Reminiscent of relationships articulated by Aristotle and Cusanus centuries earlier, modern mathematical thinking provided a new model for theological contemplations attuned to divine immanence inherent in processes and potentialities as much as to transcendental conceptions.

Chinese, Indian, and Modern Esoteric Traditions

Chinese engagements with mathematics have long been intertwined with cosmological and spiritual concerns. Astrology and divination depended on computational abilities. Consequently, one finds strong associations between mathematical practices and number mysticism, relationships not unlike those found in antiquity and throughout Europe during the Middle Ages. Even so, the desire to predict astronomical and calendrical events inspired the need to solve systems of modular congruences. Such solutions date to the thirteenth century and form the basis of the Chinese Remainder Theorem.

Mathematical practices historically associated with the Indian subcontinent also evidence spiritual influences. Ancient Vedic observances required geometric knowledge in the construction of altars that were built in various shapes with fixed areas. Similar mathematical prescriptions eventually extended to the building of temples. Vedic literature also suggests the incorporation of a symbol for zero, which became part of the Hindu-Arabic system later adopted in Europe. The symbol emerged from the considerations of Brahma as universally divine and immanent even in nothingness.

Though distinct traditions, Hinduism and Jainism attended to numerical computations as a way of contemplating the complexity and extent of the universe, including the number of ways that things might be combined. One verse from the Jainaic *sthananga sutra* (c. 300 B.C.E.) identifies algebra, geometry, and combinatorics as constituents of mathematical expertise in a way that reflects the Platonic prescription of mathematics as an essential form of knowledge.

The emergence of modern theosophy in the nineteenth century was precipitated in part by the Victorian crisis in faith and obsession with orientalism. Mathematics occupied a special place in theosophy, particularly in the numerological interests of the ancients. More contemporary concerns, however, also commanded attention within this esoteric movement. The notion of higher dimensional space, which gained credibility and notoriety through the development of algebraic methodologies and non-Euclidean geometries, was a topic of considerable discussion among theosophists. Some appealed to it by way of analogy to support beliefs in a universal present that connected the past with the future. Others made claims of brotherhood based on the notion that all of humankind is the manifestation of a single universal being that could be accommodated in an expanded conception of space. Peter Ouspensky (1878–1947) provided one of the most fulsome accounts of such thinking in his *Tertium Organum*.

Islamic Tradition

Islamic mathematics incorporated and extended ancient Greek and Indian knowledge. More significantly Muslims transmitted this expanding body of knowledge widely during the period that saw their cultural and intellectual influence spread from the Middle East to Spain (c. 700–1500). As with other cultures, astronomical considerations focused attention on geometry and trigonometry. Further, requirements associated with daily prayers, one of the Five Pillars of Islam, served to connect religious and mathematical practices. Interest in accurately establishing the five daily prayer times, which are set according to the Sun's position as determined by shadow length, provides one connection with the trigonometry of astronomical computations. Additionally, the problem of locating the direction of Mecca, toward which the faithful must face when praying, meant the Muslim mathematicians were equally concerned with the trigonometry of geography.

The significant relationships between the offering of prayers and trigonometry notwithstanding, discourses explicitly linking mathematical and theological concerns are not common features of Islamic texts dating from the Middle Ages. Patterns incorporated as architectural ornamentation may reflect natural observations rather than the realization of mathematical knowledge. However, some have suggested that the algorithmic pattern making so prevalent in Islamic architecture may reflect cosmological and theological contemplation. Specifically, it could provide a visual representation of creation that was understood in the context of number, especially in the generation of the many (numbers) from a singular unit (one). The use of multiple geometrical patterns, each integral yet distinct, may also serve a visual invitation to reflect on the parables of the Qur'an. While theological intentions might be difficult to document, mathematical expertise was certainly involved in rendering the elaborate spherical tessellations that adorn many of the domes found in Islamic architecture. Such knowledge is contained in Islamic texts such as *Those Parts of Geometry Needed by Craftsmen* (c. tenth century).

Further Reading

Ascher, Marcia. *Mathematics Is Everywhere: An Exploration of Ideas Across Cultures*. Princeton, NJ: Princeton University Press, 2002.

Henry, Granville C. *Logos: Mathematics and Christian Theology*. Lewisburg, VA: Bucknell University Press, 1976.

Hersh, Rueben. *What Is Mathematics, Really?* Oxford, England: Oxford University Press, 1997.

Koetsier, T., and L. Bergmans, eds. *Mathematics and the Divine: A Historical Study*. Amsterdam: Elsevier, 2005.

Rajagopal, Pinayur. "Indian Mathematics and the West." In *Knowledge Across Cultures: A Contribution to Dialogue Among Civilizations*. Edited by Ruth Hayhoe and Julia Pan. Hong Kong: Comparative Education Research Center, 2001.

K. G. Valente

Mathematics Genealogy Project

Category: Mathematics Culture and Identity.
Fields of Study: Communications; Connections.
Summary: The Mathematics Genealogy Project maps professional relationships among mathematicians.

Two fundamental components of the fabric of human societies are family and community. For reasons like innate socialization, sense of responsibility, and loyalty, an individual is compelled to be a part of a larger organization. In a similar way, the desire to distinguish one's place in the community, the wanting to carve out a place in the family, the urge to preserve the past for future generations, and numerous such factors motivate an individual to seek a family history. Consequently, throughout history individuals have spent much time and effort on genealogy in pursuit of their own ancestries and to reconstruct trees of ancestors.

Similarly to an individual's desire of constructing family genealogy, many professionals also have the desire and motivation to pursue their professional history. This desire is particularly the case for the professions or crafts in which some form of "apprentice" and "master" relationships are the main mode of transferring knowledge or skills from one generation to the next. Professional mathematics is a prime example of such a vocation. Particularly since the Renaissance, a prospective mathematician usually studies and conducts research under the supervision or tutelage of a master mathematician whose guidance and knowledge are major factors in obtaining successful certification to become a recognized mathematician—the Ph.D. degree.

Mathematicians usually have very high regard for this type of transfer of knowledge and profession;

Some mathematicians refer to their Ph.D. advisers as their "mathematical parents." (iStockphoto)

hence, Ph.D. advisers are given special respect. Indeed, in mathematical events, novice mathematicians' introductions typically include their adviser's name, or novice mathematicians introduce themselves as students of their adviser. Some even go as far as calling their Ph.D. adviser as their "mathematical" parent. In such an environment, it is natural for mathematicians to inquire about their mathematical ancestries. Another factor that contributes to this curiosity is, in the vastness of mathematics, finding the intertwining connections between the various subdisciplines and tracing back the original sources and motivations of the problems or concepts being studied.

Birth of the Project

The Mathematics Genealogy Project is a natural outcome of such curiosity and is the brainchild of Professor Harry B. Coonce. Although several small groups of mathematicians or some individual mathematicians had information on the genealogy of numerous prominent mathematicians, until Coonce's initial work in the late 1990s, no attempt was undertaken to construct a genealogy tree for a large group of mathematicians. In 1997, realizing that there was no central location where the information on mathematics Ph.D. students and their advisers was available, Coonce (whose adviser was Malcolm S. Robertson) started a Web site for this purpose. Upon his retirement in 1999, he devoted all his time to the project and began systematic data collection and formation of a genealogy tree for all mathematicians, which has become the Mathematics Genealogy Project (MGP). In 2003, the MGP moved to North Dakota State University (NDSU) and has been housed there since. The project's primary responsibility rests with the NDSU Department of Mathematics. In late 2009, Coonce retired from being the managing director of the project; and in October 2009, the American Mathematical Society became the sole designated partner of NDSU for MGP.

Construction of a genealogy tree is a complex process that uses historical records and other reliable sources to demonstrate kinship. Because of its unique position and its desire to provide the family tree for all mathematicians, this task is particularly difficult for the MGP. It is essentially a searchable database in which information for each entry contains all relevant professional information about that individual. The project's mission statement, quoting from the project Web site, indicates this ambitious goal clearly:

> The intent of this project is to compile information about ALL the mathematicians of the world. We earnestly solicit information from all schools who participate in the development of research level mathematics and from all individuals who may know desired information. It is our goal to list all individuals who have received a doctorate in mathematics. For each individual we plan to show the following: the complete name of the degree recipient; the name of the university which awarded the degree; the year in which the degree was awarded; the complete title of the dissertation; and the complete name(s) of the advisor(s).

In order to provide all this information as accurately as possible, the project managers gather data from reliable sources. The main sources of data are information provided from the Ph.D.-awarding institutions and the *Dissertation Abstracts*. Another important source is the mathematical community itself; voluntarily, many mathematicians provide valuable information that is not accessible to the project managers. In any case, before any entry is included in the project database, it is scrutinized for possible errors. However, some erroneous information can still be found; some of this is because of changes in the individuals' records, such as name changes because of marriage, revised spellings because of move, and name changes of institutions, and some are genuine errors. These errors are other reasons that the project administrators rely on the mathematical community for monitoring the entries and reporting and correcting the errors found.

Besides providing the information on the genealogy of mathematicians, the MGP aims to be a source of other relevant data and a hub of connections to other related projects. Therefore, the project Web site (http://genealogy.math.ndsu.nodak.edu) also contains interesting features of this kind. It provides links to databases or search tools, like MathSciNet, and links to other institutions that carry relevant information. One can also find some interesting information on the mathematicians who are most prolific and have a large number of descendants.

Mathematicians, naturally, are inclined to seek a mathematical structure within any object on which they cast their eyes. As is seen in the Extrema section of the project Web site, the MGP tree happens to have a special nonplanar graph structure. Researchers are using the data to investigate graph theoretic and visualization issues as well as social issues, like the advisers with the most students or descendants and the role of mentoring in advisee productivity. It is even possible that a new research area of mathematics on the study of structures within the MGP tree may emerge.

Further Reading

Adams, Jon. "A Trace of Greatness." *Times Higher Education* 6 (May 2010).

Jackson, Allyn. "A Labor of Love: The Mathematics Genealogy Project." *Notices of the American Mathematical Society* 54, no. 8 (2007).

Malmgren, R. Dean, Julio Ottino, and Luis Amaral. "The Role of Mentorship in Protégé Performance." *Nature* 465 (June 3, 2010).

Miller, Frederic P., Agnes F. Vandome, and John McBrewster, eds. *Mathematics Genealogy Project.* Beau-Bassin, Mauritius: Alphascript Publishing, n.d.

North Dakota State University. Department of Mathematics. "The Mathematics Genealogy Project." http://www.genealogy.ams.org.

Dogan Comez

Measurement in Society

Category: School and Society.
Fields of Study: Connections; Measurement.
Summary: Accuracy and precision are important in the many systems of measurements used in various spheres in society.

Imagine how chaotic the world would be if people could not measure anything! People would not be able to keep track of time, would not know weights or heights of people (or of anything else in the world), could not calculate the distance between any two points, and would not have recipes to cook properly. Indeed, the list of everyday activities that would be impossible to do in the absence of measurement is endless. Thus, measurement is an essential part of everyday life. Measurement is a fundamental part of mathematics research and curricula and there are many types of measurements in society. Some measurements elucidate productivity or change. Others measure large-scale aspects of society, like gross domestic product (GDP). Area measurements have practical applications in areas like surveying and interior design. Measurements are fundamental in drug dosing labels, quality control, missile launches, and in many other applications and fields. Because of its critical and practical importance, measurement is an extensively studied concept in pre-K–12 mathematics education.

Measurement Systems

Numerous measurement systems have been developed and used since ancient times, the earliest of which used body parts as the unit of measurement. The many, diverse measurement systems were a source of confusion, not only among nations, but also among different fields within a nation. To establish common units of measurement and promote their use, a treaty titled the "Convention of the Metre" was signed by 17 countries on May 20, 1875. The Convention of the Metre established three international organizations—the International Bureau of Weights and Measures (BIPM), the General Conference on Weights and Measures, and the International Committee for Weights and Measures—to oversee issues related to measurement in the member nations.

In 1960, the 11th General Conference on Weights and Measures developed and adopted a unified measurement system named International System of Units (SI) to promote a worldwide measurement system. The SI is based on seven dimensionally independent units: meter (the unit of length; abbreviated as m), kilogram (the unit of mass; abbreviated as kg), second (the unit of time; abbreviated as s), ampere (the unit of electric current; abbreviated as A), kelvin (the unit of thermodynamic temperature; abbreviated as K), mole (the unit for amount of substance; abbreviated as mol), and candela (the unit of luminous intensity; abbreviated as cd). Although the spelling of the base units may differ in different languages, the symbols are the same worldwide. The SI is an evolving measurement system to keep up with ever-growing measurement needs. The BIPM, which is comprised of many countries, ensures that measurements throughout the world are traceable to the SI. The BIPM is related to other significant international organizations such as the International Commission on Illumination, the International Atomic Energy Agency, the International Laboratory Accreditation Cooperation, the World Health Organization, and the International Organization of Standardization. Such a worldwide organization to oversee the uniformity of measurements explains clearly the reason for the crucial emphasis on measurement in mathematics curricula.

The United States has its national standards for measurement and measuring devices explained in the U.S. Code. Because measurement and measurement devices are a part of everyday life and are used in various businesses and for commercial purposes, the U.S. Code, published by the Office of the Law Revision of

---

Definitions of the Seven Base Units of SI

Length
meter, m: The meter is the length of the path travelled by light in vacuum during a time interval of 1/299,792,458 of a second.

It follows that the speed of light in vacuum, C_0, is 299,792,458 m/s exactly.

Mass
kilogram, kg: The kilogram is the unit of mass; it is equal to the mass of the international prototype of the kilogram.

It follows that the mass of the international prototype of the kilogram, m(K), is always 1 kg exactly.

Time
second, s: The second is the duration of 9,192,631,770 periods of the radiation corresponding to the transition between the two hyperfine levels of the ground state of the cesium 133 atom.

It follows that the hyperfine splitting in the ground state of the cesium 133 atom, v(hfs Cs), is 9,192,631,770 Hz exactly.

Electric Current
ampere, A: The ampere is that constant current that, if maintained in two straight parallel conductors of infinite length, of negligible circular cross-section, and placed 1 meter apart in a vacuum, would produce between these conductors a force equal to 2×10^{-7} newton per meter of length.

It follows that the magnetic constant, μ_0, also known as the permeability of free space, is $4\pi \times 10^{-7}$ H/m exactly.

Thermodynamic Temperature
kelvin, K: The kelvin, unit of thermodynamic temperature, is the fraction 1/273.16 of the thermodynamic temperature of the triple point of water.

It follows that the thermodynamic temperature of the triple point of water, T_{tpw}, is 273.16 K exactly.

Amount of Substance
mole, mol:
1. The mole is the amount of substance of a system that contains as many elementary entities as there are atoms in 0.012 kilogram of carbon 12.
2. When the mole is used, the elementary entities must be specified and may be atoms, molecules, ions, electrons, other particles, or specified groups of such particles.

It follows that the molar mass of carbon 12, M(^{12}C), is 12 g/mol exactly.

Luminous Intensity
candela, cd: The candela is the luminous intensity, in a given direction, of a source that emits monochromatic radiation of frequency 540×10^{12} hertz and that has a radiant intensity in that direction of 1/683 watt per steradian.

It follows that the spectral luminous efficacy, K, for monochromatic radiation of frequency 540×10^{12} Hz is 683 lm/W exactly.

(Adapted from the Bureau International des Poids et Mesures (BIPM) Web site at http://www.bipm.org/utils/common/pdf/si_summary_en.pdf)

the House of Representatives, includes a chapter titled "Weights and Measures and Standard Time" under the Title 15. The chapter sets standards for weight and measurement devices to enforce accuracy and to ensure equity in the marketplace. In the early twenty-first century, the acknowledgement of measurement in the U.S. Code as a chapter containing 267 sections under nine subchapters is a sound indicator of the importance of measurement in human life.

Accuracy and Precision in Measurement

Any measurement is an approximation to the real value of a quantity. The length of the previous sentence might be 12 centimeters (cm), but in millimeters (mm) it would be 121 mm; 12 cm is not equal to 121 mm. The reason behind the difference between these two measurements is the second measurement is more accurate than the first one. Can it be measured more accurately? This question yields to the need for accurate measurement. One millimeter in this example can be ignored, but an inaccuracy of a mere millimeter in a missile launch may result in a disaster. Improvements in measurement systems are extremely important to make measurements as accurate as possible.

In measurement, the most accurate and precise results are desired. "Accuracy" in measurement refers to the extent to which a measured value matches the correct value. "Precision," on the other hand, refers to the reliability of a measurement and how close individual measurements are to each other. Measurement units and devices in different fields of study are not static; rather, they evolve to improve accuracy and precision. In the United States, the National Institute of Standards and Technology (NIST) is a federal agency that employs mathematicians and scientists, among others, whose main tasks include the advancement of the science of measurement and measurement standards. NIST, together with partners from the government, industry, and academia, also develops measurement tools for different sciences. The services of NIST include verification of the accuracy of measurements, instrument calibration (for example, calibration of dimensional, mechanical, or electromagnetic instruments) to improve measurement quality, and the development of innovative measurement methods.

Although accuracy and precision are always desirable in measurement, in some fields quality of measurement is more crucial. For example, the National Aeronautics and Space Administration (NASA) uses various instruments to measure temperature, pressure, load, and acceleration, and to make other critical measurements for its test programs. The Measurement Standards and Calibration Laboratory of the White Sands Test Facility, which supports an extensive number of test programs, performs instrument calibrations to ensure measurement quality is compatible with recognized national standards that are traceable to NIST. In NASA's test programs, any error in measurements in any equipment may cause not only the deaths of highly trained astronauts but also the loss of millions of dollars. Accurate and precise measurement therefore underpins the success of NASA missions, including launching spaceships and ensuring their safe return to Earth.

Another field where measurement accuracy has critical importance is the health industry. Cancer is one of the most serious diseases that the human race has faced so far. Almost 13% of all deaths in the world were caused by cancer in 2004. Radiotherapy, which uses high-energy radiation to kill cancer cells, is one of the most frequent methods used to treat cancer patients. However, radiotherapy not only kills cancer cells but kills healthy cells as well. Before the start of a cancer treatment, doctors conduct a simulation to locate the patient's tumor and the normal tissues around it. In order to provide effective treatment, the next step is to measure the dose of the radiotherapy required and the safest angles to deliver the radiation to kill cancer cells. Measurements taken for radiotherapy have to be as precise as possible because the amount of radiation required to kill a cancer cell differs by the type of cancer cell and there is a risk of damaging the normal tissue during the radiation delivery. Sophisticated computers capable of making sensitive measurements are used for radiotherapy planning. Thus, human error in measurement is decreased. Advancements in technical equipment used in cancer treatment help to increase the effectiveness of the treatment and decrease the deaths caused by cancer. Monitoring the patient's temperature and thermal dosage in real time provides doctors the opportunity to treat tumors as closely as possible while keeping the adjacent healthy tissues safe.

Measurement in Everyday Life

Measurement is a pervasive mathematical concept in everyday life, so it has many applications to a variety of careers, such as health sciences, architecture and construction, interior design, carpentry, meteorology, and public safety. Precise measurement is crucial in healthcare, as monitoring patient condition has critical importance. Thus, choosing effective measurement devices and obtaining accurate measurements (for example, of weight, blood pressure, or blood sugar) are essential aspects of healthcare professions. Also, healthcare professionals frequently use measurement conversion on the job. Doctors, nurses, and pharmacists convert between English and metric systems, or between Celsius and Fahrenheit, when they collect patient information on weight or temperature or when calculating appropriate medication dosages to administer. Measurement conversion is a particularly important competency for pharmacists, as they convert among different measurement systems such as metric, apothecary, and avoirdupois systems when they calculate medication dosages and fill orders.

Measurement is among the essential mathematics concepts applied in architecture, construction, and related careers. From the design and scale models of a project to its actual construction, precise and accurate measurement is vital. Measurement is also used extensively by interior designers as they improve the aesthetics and function of interior spaces. Interior designers have to determine precise measures of virtually all parts of a space to most effectively utilize the space and to decide the type, size, and placement of furniture or fixtures. Designers need to have precise area measures of walls, floors, or countertops to determine the size and number of tiles needed to cover these surfaces. Indeed, site measure and survey is an essential routine for interior designers in which they get measures of a space and draw an outline of the space, including dimensions.

Carpentry is another occupation for which measurement is substantially important. An old saying emphasizes the significance of measurement in carpentry: "Measure twice, cut once." Because precise measurement is at the heart of good carpentry work, carpenters use various specialized measurement tools, such as a combination square (to accurately measure 45 degree and 90 degree angles), carpenter's square (to plot right angles), and T-bevel (to set and transfer angles), in addition to the regular metal tape measures and folding rulers.

Although most people are familiar with thermometers and their uses, many may not know about various other measurement scales meteorologists use to organize and record weather conditions. Meteorologists use anemometers to measure wind speed or pressure, ceilometers to measure the thickness and height of clouds, barometers to measure atmospheric pressure, and high-tech sensors to measure humidity. People have always been interested in reliable and long-term weather forecasts. Although weather predictions are increasingly accurate and can be made for increasingly longer terms, meteorologists are continuously searching for methods to improve weather predictions. In this effort, innovative measurement devices in meteorology are being developed using the most up-to-date technology to make more accurate and precise weather and climate predictions.

Measurement also has significant applications in public safety. To maintain public travel safety, the Transportation Security Administration (TSA) utilizes the most advanced imaging technology, such as millimeter wave scanners to screen passengers for metallic and nonmetallic threats that might be anywhere on the body without physical contact. Millimeter wave scanners use electromagnetic waves to produce a black-and-white image in seconds. These scanners transmit extremely high radio frequencies, a wavelength of 1–10 mm, from two antennas to construct a three-dimensional image of the person scanned. The energy each radio wave reflects back from the passenger's body to the scanner is transmitted to a computer. Then, software measures the energy for each radio wave reflected from the passenger's body to construct an accurate and precise three-dimensional image of the passenger for security check. With the help of such detailed three-dimensional images, any hidden object can easily be identified by security. For such an imaging technology to be used in areas requiring high security needs, like airports, the technology needs to provide fast, accurate, and reliable images. Further, imaging technology developers should consider the amount of radiation emitted by a person who is screened. With more accurate and reliable measurements using advanced imaging technologies, human life can be protected both by eliminating possible threats to public safety and by decreasing side effects of such screening technologies.

Measurement in Pre-K–12 Mathematics Curricula

The study of measurement starts before kindergarten, and most children of pre-K and kindergarten age can acquire considerable knowledge of measurement. Providing young children with motivating opportunities to explore measurable characteristics of objects such as size, weight, and length and engaging them in activities that require comparing and ordering objects by these characteristics can help them develop the concept of measurement. For example, children can order their toys by their size, make short and long (or big and small) animals using clay, or match items of the same size. An activity that can help children start developing an understanding of area might be covering a large flat surface using small sizes of the same surface (such as leaves or cookies) and making comparisons between surface areas (for example, a larger leaf or a smaller cookie). Children can develop a general idea of volume as they pour water from a wider to a narrower container, or from a taller to a shorter container. Parents can also contribute to their children's learning of early measurement concepts and appropriate measurement terms by making comparisons using terms such as "big," "bigger," "small," "smaller," "light," "lighter," "heavy," "heavier," "tall," "taller," "short," and "shorter" when referring to objects or people in their daily conversations. In their daily routines, children encounter various opportunities to develop an understanding of time and its measurement. For example, children can understand the day and night cycle and sequences of their daily activities (washing hands before meals and brushing teeth after meals). The waiting periods for major events that children look forward to, such as special days and holidays, can provide opportunities for children to understand concepts of day, week, month, and year. Young children can learn various measurement devices within daily contexts as they associate money with buying things, clocks and calendars with time concepts, or thermometers with temperature.

In addition to making comparisons and ordering familiar objects, children should experience the process of measurement. Before being introduced to standard units of measure, such as inches or feet (or equivalent units in the metric system), children typically start measuring using nonstandard measurement units. For linear measurement children can measure the length of a table using their hands, the height of a chair using paper clips, or the distance between two points using their feet. Children can explore measuring area as they cover different sizes of flat objects with uniform blocks. An activity for children to learn about volume measurement is placing uniform cubes in a box and counting the number of cubes used to fill up the box. Balance scales can be used to provide children with comparisons of weights of different objects, such as comparing an eraser's weight to a pencil's weight. Children can also weigh objects with nonstandard units using balance scales. They can weigh a book using unifix blocks or a pencil using paper clips. The Illuminations Web site of the National Council of Teachers of Mathematics (NCTM) provides various lesson samples that can be used in preschool classrooms or at home to teach children measuring with nonstandard units. When measuring with nonstandard units, students can conceptualize that they determine the total length, area, volume, or any attribute of interest as they repeatedly measure using the same measurement unit.

After children experience the measurement process using nonstandard units, they will be better prepared to explore measuring with standard measurement tools and units. Measuring with standard units as well as nonstandard units is among NCTM's measurement standards for grades pre-K–12. According to NCTM standards, pre-K–12 students also should be able to choose appropriate measurement units and tools to measure different attributes. Students can be introduced to standard measurement tools, such as tape measures, scales, or rulers, with activities that allow them to experiment with the measurement process. For

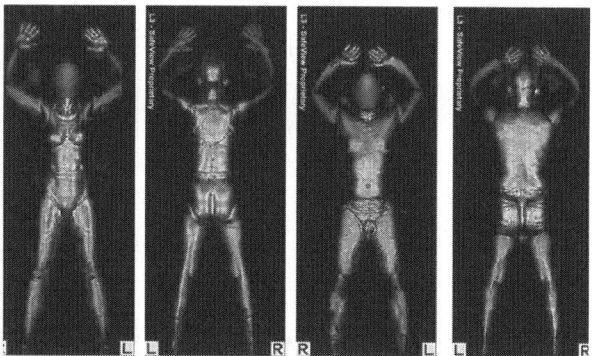

The U.S. government claims that millimeter wave scanners emit less energy than a cell phone. (Transportation Security Administration)

example, to learn measuring weight using a scale and to gain an idea about weights of different objects, students can weigh themselves and various items such as a bag, a book, or fruit on a scale and record the weights. As students weigh using the scale, they will recognize the units of measurement. After students gain some experience with measuring weights, an enjoyable activity might be to ask students to estimate weights of things that they identify in the classroom.

Throughout elementary and middle school, students learn conversions within a measurement system; measure time, area, volume, temperature, and angle size using appropriate measurement units and tools; find the areas of rectangles, triangles, parallelograms, circles, and irregular shapes; and calculate volumes and surface areas of rectangular solids, cylinders, and trapezoids. In later grades, students are expected to analyze measurement precision and accuracy and approximate measurement error.

An important concept that students need to learn when they study measurement is "estimation." Students in early grades can determine common or personal referents (for example, the width of an index finger is 1 centimeter) as they estimate different attributes, such as length and weight of common objects. As students move on to higher grades, they should be prompted to estimate perimeters, areas, and volumes using benchmarks. Students in college explore the theory of measurement. For instance, they use techniques from calculus to represent the length of a curve and the notion of a metric space is defined in topology. Mathematicians measure hard-to-define quantities, like the length of a coastline, refine and improve systems of measurement, and also research related concepts in the field of measure theory.

Further Reading

Confer, Chris. *Sizing Up Measurement: Activities for Grades 3–5 Classrooms.* Sausalito, CA: Math Solutions Publications, 2007.

Drum, Randell L., and Wesley G. Petty, Jr. "2 Is Not the Same as 2.0!" *Mathematics Teaching in the Middle School* 6, no. 1 (2000).

Howarth, Preben, and Fiona Redgrave. *Metrology: In Short.* 3rd ed. Albertslund, Denmark: Schultz Grafisk, 2008.

National Council of Teachers of Mathematics. "Illuminations: Resources for Teaching Math." http://illuminations.nctm.org.

National Council of Teachers of Mathematics. *Principles and Standards for School Mathematics.* Reston, VA: National Council of Teachers of Mathematics, 2000.

Nowlin, Donald. "Precision: The Neglected Part of the Measurement Standard." *Mathematics Teaching in the Middle School* 100, no. 5 (2007).

Zeynep Ebrar Yetkiner Ozel
Serkan Ozel

Musical Theater

Category: Arts, Music, and Entertainment.
Fields of Study: Communication; Geometry; Number and Operations; Representations.
Summary: Mathematical concepts and mathematicians have become interesting subjects of musical theater.

After the popular and critical success of Tom Stoppard's *Arcadia*, first performed in London in 1993, playwrights began making regular use of mathematics as source material for new scripts. This interdisciplinary collaboration, however, has largely been confined to stage plays and in the early twenty-first century has not found its way into musical theater—with one glaring and quite remarkable exception. In 2000, the husband-and-wife team of Joanne Sydney Lessner and Joshua Rosenblum created *Fermat's Last Tango*, a comic musical inspired by Princeton mathematician Andrew Wiles and his successful proof of Fermat's Last Theorem.

Fermat's Last Tango

Fermat's Last Theorem (FLT) is arguably the most famous mathematical problem in history. When Pierre de Fermat left his tantalizing note in the margin of his copy of Diophantus' *Arithmetica* in 1637, the result was a challenge that resisted the efforts of mathematicians for the next 350 years. By the twentieth century, FLT had acquired such a daunting reputation that when Princeton mathematician Andrew Wiles decided to take it on around 1986, he did not tell anyone what he was doing until seven years later, when he emerged from the office in his attic with what he thought was a proper proof of the Taniyama–Shimura conjecture. A proof of Taniyama–Shimura was known to imply FLT, and the unassuming Wiles was propelled to unprecedented stardom far beyond the mathematical community.

This event is the jumping-off point for *Fermat's Last Tango*. Because of the fictional liberties they take with the story, Lessner and Rosenblum have changed the name of their protagonist from Andrew Wiles to Daniel Keane, and the first major piece of revisionism we experience is when Keane is visited by a devilish and vindictive Fermat and whisked off to "the Aftermath" to fraternize with Pythagoras of Samos, Euclid of Alexandria, Isaac Newton, and Carl Friedrich Gauss. The fantasy is enjoyable, but what is really striking is how few liberties are taken with the mathematics. That the authors have done their homework is clear early on when Fermat rhymes "Shimura–Taniyama" with "algebraic melodrama." In the Aftermath, Fermat reveals that Keane has made some incorrect assumptions about the Galois representations he used in his argument—which is indeed a mistake Wiles had made—and Keane retreats to his attic to try to repair the "big fat hole" in his proof.

Wiles, like Keane, was deeply uncomfortable trying to fill the gap in his proof under the glare of public scrutiny. The writers also keep the touching anecdote that Wiles promised his wife a corrected proof by her birthday, although it is unlikely that the real Ms. Wiles tried to lure her husband away from his research by crooning "Check out my modular form." Taken in the lighthearted spirit in which it was intended, *Fermat's Last Tango* is roundly successful entertainment. Beyond this achievement, it also comes as close as any other piece of science theater to effectively staging the "moment of discovery," creating a genuinely breathless moment when a defeated Keane finally realizes how to repair the hole in his proof using the Iwasawa theory approach he had abandoned several years earlier.

For those who do not have an opportunity to view a live production, a performance of *Fermat's Last Tango* was recorded and is available through the Clay Mathematics Institute. Others have staged *Fermat's Last Tango* specifically as a teaching experience. A 2007 article in *PRIMUS*, a publication dedicated to teaching undergraduate mathematics, describes a fully student-mounted production, along with suggestions for related educational activities. The play is cited as a good introduction to not only mathematics products but also the personalities of people and the processes involved in mathematics research.

The Natural Sciences

There does not seem to be any other piece of widely disseminated musical theater devoted to a mathematical topic, though certainly there are mentions of mathematics in various popular scores. For instance, in *Pirates of Penzance*, first performed in 1879, W. S. Gilbert and Arthur Sullivan include the following stanza in the famously tongue-twisting Major-General's song:

> I'm very well acquainted, too, with matters mathematical
> I understand equations, both the simple and quadratical
> About binomial theorem I'm teeming with a lot o' news
> With many cheerful facts about the square of the hypotenuse

Broadening the net to include the mathematical sciences brings into play the work of American composer Philip Glass. In 1976, Glass scored and wrote *Einstein on the Beach*, which was viewed as groundbreaking in several ways—one being that it was nearly five hours long with no intermission. The implication here was that audience members were expected to come and go as they so desired. In a similar vein, it was not plot driven but did contain many references to Einstein, including a musical event meant to suggest a nuclear explosion.

In 2001, Glass wrote the music for the opera *Galileo Galilei*, which tells the life story of Galileo in reverse. The opera opens with Galileo blind and on his deathbed, follows him back through his trial and astronomical discoveries, and ends with Galileo as a child attending an opera written by his father. Glass returned to the natural sciences a third time in 2010 when he wrote the music for *Kepler*, an opera that features Johannes Kepler as the only named character, although there are six other soloists and a chorus.

Glass did study mathematics early in his education before devoting himself wholly to music, and he readily admits to seeing mathematics and music as being linked—not just technically but artistically. "The beauty of mathematics is something that mathematicians talk about all the time," Glass said in a November 2009 feature for the *Wall Street Journal*. "And the elegance of a mathematical theorem is almost as good as its proof. Not only is it true, but it's elegant. So you get into almost aesthetic questions."

Kepler and Galileo are also the featured characters in a 2001 musical called *Star Messengers*, written by Paul Zimet with music composed by Ellen Maddow. A much more widely toured musical production was *Dr. Atomic*, written by John Adams with libretto by Peter Sellars. This opera tells the story of the Manhattan Project largely through the eyes of physicist J. Robert Oppenheimer, in part by borrowing text from government documents and interviews with scientists who worked on the bomb. First produced in San Francisco in 2005, *Dr. Atomic* has since been performed at multiple locations in Europe and the United States, including a live broadcast from the Metropolitan Opera in 2008.

Further Reading

Chin, Cynthia E. "Mathematical Heroes—No Longer Unsung." *PRIMUS* 17, no. 1 (2007).

Shepherd-Barr, Kirsten. *Science on Stage*. Princeton, NJ: Princeton University Press, 2006.

Stephen Abbott

Numbers and God

Category: Friendship, Romance, and Religion.
Fields of Study: Number and Operations; Representations.
Summary: Many numbers and mathematical ratios are associated with religion and the notion of deity.

Numbers and religion have been linked since at least the beginning of recorded history. Many societies throughout the world have associated numbers with their spiritual beliefs. Some of these numbers still play a role in the fabric of society's belief systems, religious rituals, artistic renderings, and symbolisms. They continue to be explored, evaluated, and recognized in the religious teachings and traditions of many of the world's religions. As early as 1150 B.C.E., Indian mathematician Bhaskaracharya attributed the creation of the base-10 numeration system and zero to the Hindu god Brahma. Many ancient cultures and societies believed that certain numbers had spiritual significance. Historians, mathematicians, religious scholars, and others interested in such connections have found evidence of such beliefs in civilizations and religions like ancient Babylonia, the Society of Pythagoreans, Greece, Hellenistic Alexandria, Judaism, Christianity, and Islam. Many of these same beliefs continue into the twenty-first century.

Numbers of Pythagoras

Pythagoras of Samos (570–495 B.C.E.), who is often called the first pure mathematician, and his followers, the Pythagoreans, are well-known for their mathematical, philosophical, and religious beliefs. In antiquity, philosophy was believed by many to encompass the very essence of mathematics and religion. The perceived link between mathematics and the spiritual or divine world is succinctly stated by the Pythagoras maxim "All is Number." Among the legacies associated with Pythagoras are the theorem that bears his name; the creation and study of musical harmonies, which may have originated in Babylon; and concepts of sacred geometry, such as the divine proportion.

The "Divine Proportion" (or "Golden Ratio") is often seen by mathematicians and other scholars in nature's designs and natural phenomena. The Greeks widely used the principle in sculptures and architecture. Phidias (490–432 B.C.E.), who is counted among the best Greek sculptors, used the Divine Proportion in designing the Parthenon, a temple to the goddess Athena. In honor of Phidias, the Divine Proportion is usually symbolized by the Greek letter φ representing the first letter of his name. To understand the Divine Proportion, consider a rectangle. The rectangle is said to be in Divine Proportion if the ratio of its length to its width has the following value

$$\varphi = \frac{1+\sqrt{5}}{2} \approx 1.618.$$

Rectangles with these proportions are called "Golden Rectangles." This proportion (about 8/5) continues to be used by artists and architects in designing structures for aesthetic appeal.

The Parthenon was built in Athenian Acropolis, Greece, to honor the Greek goddess Athena. It is thought to be the perfect example of a Doric temple and was designed using the **Divine Proportion.** *(Photos.com)*

The Number "12"

The Greeks considered the number "12" to be significant since it represented the number of gods on Mt. Olympus: Zeus, Hera, Athena, Poseidon, Apollo, Artemis, Demeter, Hermes, Aphrodite, Ares, Hephaestus, and Hestia. The significance of "12" probably originated with the Sumerians in Mesopotamia. Later, the Babylonians used the number "12" in developing their calendars and their clocks. They developed the zodiac by dividing the heavens into 12 equal sections named for constellations, one for each calendar month. These sections continue to be the 12 signs of the zodiac, an idea that was passed down from society to society throughout the ages. The Babylonian zodiac impacted many societies in the Western world. In Christianity, the 12 disciples of Jesus are usually considered to be symbolic of the 12 tribes of Israel, which may have been influenced by the 12 signs of the zodiac. The number "12" also has significance in Buddhism. For example, the Buddhist Wheel of Life, which depicts the world and the human condition, has 12 stages. In this tradition, life is composed of 12 stages, which keep the wheel of life turning.

The Number "7"

The number "7" is a significant number in Judeo-Christian and Islamic religious traditions. The creation story in the book of Genesis states that God made the heavens and the Earth in six days and rested on the seventh day. The number "7" is associated with divine completion and perfection. There are also references to 7 spirits, 7 churches, 7 stars, 7 seals, 7 trumpets, 7 vials, 7 thunders, 7 plagues, 7 mountains, and 7 kings, and many more references. The number "7" occurs frequently in Muslim architecture, art, and literature. The Qur'an often couples the number "7" with references to Allah as the all-powerful creator as well as with concepts like the 7 heavens, the 7 Sleepers of Ephesus, and the 7 periods of creation.

The Number "19"

In 1974, Rashad Khalifa used a computer to explore the structure of the Qur'an. He discovered that the number

19 occurred with unusual frequency. This occurrence was unexpected since 19 had never before been recognized as a significant number in the Islamic religion. Khalifa published his discovery in his 1981 work, *The Computer Speaks: God's Message to the World*. These findings were called the "Qur'an Code." The first verse of the Qur'an states: "In the name of Allah, the compassionate, the merciful." In Arabic, the letters that make up this verse total 19. Khalifa discovered that every word in this verse is mentioned a number of times throughout the Qu'ran, and these numbers are all multiples of 19. Consequently, Khalifa's conclusion was that the number "19" was divinely selected as a number of significance in the Islamic religion.

Bible Codes

What are often now known as "bible codes" were popularized in the twentieth century, but numerical symbolism dates back to much earlier times. The Jewish book *Sefer Yetzira* (Book of Creation) contained sacred numbers. As writer and scientist Clifford Pickover has explained: "Kabala is based on a complicated number mysticism whereby the primordial One divides itself into 10 sephiroth [numbers] which are mysteriously connected with each other and work together. 22 letters of the Hebrew alphabet are bridges between them." In gematria methods of analysis, each letter was assigned a number. The values of a word or phrase were added and the then values were analyzed for spiritual implications. For example, the word for "life" in Hebrew is *chai*, which is made up of two letters, a *chet* (8) and a *yud* (10). When added together, they sum to 18. The number 18 then took on symbolic meaning, which also translated to daily life. It was considered good form to give monetary gifts in 18 and its multiples. The number 18 has been also considered prosperous in certain parts of China, and it also took on spiritual importance in India, such as in the 18 chapters of the sacred Hindu text Bhagavad Gita.

Researchers have mathematically examined the Bible using methods such as two-dimensional arrays, which have been tested for what are known as "equidistant letter sequences." Some found what seem to be words meaningfully related to adjacent portions of the text, and they claimed that their results were statistically unlikely to be due to chance alone. Author Michael Drosnin reported on some mathematical and computer analyses, referring to them as the "Bible code," in order to highlight apparent predictions and to compare to twentieth-century knowledge. Some of the advocates of Bible code analyses point to apparent prediction of the dates of major world events as proof of the existence of such codes. Computations on the age of the universe are also sometimes cited as evidence, such as when first century rabbi Nechunya ben Hakanah used the Bible to compute the age as 15.3 billion years, which is relatively close to some twenty-first-century estimates. Critics, however, have countered these assertions by citing flaws in the statistical methodology and noting that any sufficiently long text may produce seemingly nonrandom patterns or clusters.

Numerical Defense of the Resurrection

During the twenty-first century, associations of numbers with religion continue to evolve. In 2002, Richard Swinburne, philosophy professor at Oxford University, applied Bayesian statistical methods, named for mathematician Thomas Bayes, in his defense of the Christian tenet of Jesus's resurrection from the dead. He noted that it was extremely improbable, based on the laws of nature, for someone who had been dead for 36 hours to come back to life. Swinburne asserted that if there is a God, only God would be able to defy the laws of nature and make the dead come alive. In proving his point, Swinburne assigned probability values to the existence of God and some of the events described in the New Testament, such as the credibility of witness testimony. After mathematical analysis, Swinburne concluded that Jesus's resurrection was extremely probable, namely, 97 percent. Swinburne's use of mathematical logic and statistical methods to answer questions of faith is another step in a long tradition of connections between numbers and religion.

Further Reading

Brooke, John Hedley and Ronald L. Numbers. *Science and Religion Around the World*. New York: Oxford University Press, 2001.

Dudley, Underwood. *Numerology, or, What Pythagoras Wrought*. Washington, DC: Mathematical Association of America, 1997.

Pickover, Clifford. *The Loom of God: Tapestries of Mathematics and Mysticism*. Reprint. New York: Sterling Publishing, 2009.

Voss, Sarah. *What Number Is God?: Metaphors, Metaphysics, Metamathematics, and the Nature of Things*. Albany: State University of New York Press, 1995.

Sharon Whitton

Origami

Category: Arts, Music, and Entertainment.
Fields of Study: Geometry; Representations.
Summary: People explore many interesting mathematical questions using the art and principles of paper folding.

Origami is the famous Japanese art of paper folding. Historically, it has been used for a variety of purposes, including document certification and as a way to represent religious symbols. In traditional origami, a single piece of paper is folded to construct one of a variety of objects. The art has grown to include compound forms that involve connecting several individual origami pieces together, with modular origami specifying geometrically equal pieces.

Origami art, mathematics, and science have many explicit interconnections, and in the 1990s and 2000s there have been several conferences specifically devoted to these links. In the twenty-first century, computational origami is an emerging discipline that applies mathematical theory and computational algorithms to formulate and solve complex folding problems, many of which have applications in engineering, industrial design, and a variety of sciences. Such solutions are often called "origami technology." For example, engineers and mathematicians explored origami lenses for use in space telescopes, and precision folding technology is already being used to optimize manufacturing processes.

Origami forms are inherently mathematical. Their geometry can be identified as reflections with respect to the folding line. The possible operations for points and lines in origami, using a single fold, are described by seven axioms generally known as the "Huzita–Hatori axioms," named for mathematicians Humiaki Huzita and Koshiro Hatori. However, mathematician Jacques Justin may have been the first to enumerate these seven axioms. The axioms allow mathematicians to answer interesting questions, such as the classic problems of trisecting an angle and doubling the cube, which are impossible using only ruler and compass constructions. More generally, it is possible to solve any equation up to degree three with origami geometry. Further, although origami forms are usually produced using finite sheets of paper, origami folding can theoretically be extended to the infinite plane.

Use of Origami in Modern Mathematics

In the late twentieth century, mathematicians got interested in the foundations of this art. For this community of scientists, the creation of models in origami is not a matter of inspiration; it is spurred by the search for understanding of the concepts and limitations of Euclidean geometry, properties of geometric figures, symmetry, angles, lines, and mathematical communication, among others.

There are several major topics in the practice and study of origami, including the following:

- Its geometry and relationship between this and other geometries, in particular, Euclidean geometry
- The straightening of the bend—whether a model can be unfolded (which has been studied by Marshall Bern and Barry Hayes)
- Rigid origami—the possibility of constructing models if the paper were replaced by metal (which has already been used for solar panels of satellites in space)

Mathematics teaching techniques increasingly use origami. Moreover, paper folding is used to develop manual dexterity, as well as to teach aesthetics appreciation and topics such as proportions, foundations of geometry, and measurements. Origami is also a handy resource for other areas, like mathematical communication, problem solving, and investigation of three-dimensional objects and spatial relationships.

Huzita–Hatori Axioms

1. Given two points P_1 and P_2, we can fold a line connecting them.
2. Given two points P_1 and P_2, we can fold P_1 onto P_2.
3. Given two lines l_1 and l_2, we can fold line l_1 onto l_2.

4. Given a point *P* and a line *l*, we can make a fold perpendicular to *l* passing through the point *P*.
5. Given two points P_1 and P_2 and a line *l*, if the distance between P_1 and P_2 is equal to or larger than the distance between P_2 and *l*, we can make a fold that places P_1 onto *l* and passes through the point P_2.
6. Given two points P_1 and P_2 and two lines l_1 and l_2, if the lines aren't parallel and if the distance between the lines isn't larger than the distance between the points, we can make a fold that places P_1 onto line l_1 and places P_2 onto line l_2.
7. Given a point *P* and two lines l_1 and l_2, if the lines aren't parallel, we can make a fold perpendicular to l_2 that places *P* onto line l_1.

Robert Lang proved that this list of axioms covers all possible cases for a single folding. If one of them is removed from the list, it is no longer complete.

Further Reading

Demaine, Erik, and Joseph O'Rourke. *Geometric Folding Algorithms: Linkages, Origami, Polyhedra*. Cambridge, England: Cambridge University Press, 2007.

Hull, Thomas C. *Project Origami: Activities for Exploring Mathematics*. Wellesly, MA: A K Peters, 2006.

Lang, Robert J. "Origami and Geometric Constructions." http://www.langorigami.com.

———. *Origami Design Secrets: Mathematical Methods for an Ancient Art*. Natick, MA: A K Peters, 2003.

Liliana Monteiro

Painting

Category: Arts, Music, and Entertainment.
Fields of Study: Geometry; Representations.
Summary: Painting incorporates many mathematical concepts, and mathematics is also used to analyze paintings.

Human beings strive to comprehend their reality in a number of ways, including artistic expression and mathematics. Examples can be found in many cultures, such as the long history of interesting mathematical patterns in Islamic art and in the cave paintings of Paleolithic people. Many artists throughout history also have been mathematicians, such as fifteenth-century painter Piero della Francesca. Modern painter Michael Schultheis also worked as a software engineer. He and Mary Lesser, a painter and printmaker, both explicitly include mathematical elements like numbers, equations, and geometric objects in their work. Mathematical concepts, especially geometry, are embedded throughout the art of painting. Some that are most commonly used for analyzing paintings involve symmetry, perspective, golden ratios and rectangles, and fractals, as well as fundamental geometric forms, shapes, fractals, and abstraction. Mathematicians and scientists also use mathematical methods to determine whether or not unidentified paintings belong to a particular artist.

Symmetry

M.C. Escher, a graphic artist, used transformational geometry to create a variety of works that explored symmetry. His classic work *Day and Night*, a 1938 woodcut, transforms rectangular fields into flying geese and uses a black and white color scheme to emphasize the transition of a setting from day to night. While many artists explore symmetry and transformational geometry, Escher took it further by exploring and emphasizing mathematical concepts including *Convex and Concave*, a 1955 lithograph, *Two Intersecting Planes*, a 1952 woodcut, and *Moebius Strip II*, a 1963 woodcut. The use of symmetry as the catalyst for transforming the plane is one of the more pleasing aspects of his work. Navajo sand painting also offers many good examples of various types of symmetry. Four-fold symmetry is widely found in Native-American painting and other art forms, and it plays a role in some spiritual and healing ceremonies.

Perspective

Early paintings did not use perspective to show a three-dimensional world on a two-dimensional canvas. Giotto di Bondone, a thirteenth-century painter began to develop depth of field in some of his work; but the first artist credited with a correct representation of linear perspective is Filippo Brunelleschi (1377–1446), who was able to devise a method using a single vanishing point. An architect and sculptor, he shared his method with fellow artist Battista Alberti, who wrote about the mechanics of mathematical perspective in painting. Leonardo da Vinci used perspective in his paintings and explored artificial, natural, and compound perspective in his work. He examined how the viewer's observation point changed the perspective, and how the perspective could be perceived by changing where the viewer was observing the painting. Notably, while perspective and the illusion of depth were widely used in Western painting from the 1300s onward, it was not universal. Painters from India rarely used this technique; rather, they tended to focus more on patterns and geometric relationships.

French painter Georges Seurat used the painting technique of pointillism to create Sunday Afternoon on the Island of La Grande Jatte. *(Yorck Project)*

Golden Ratio and Golden Rectangles

Consider a rectangle with short side a and a long side that is $a+b$. A golden rectangle would be where the ratio a/b is equal to

$$\frac{a+b}{a}.$$

In other words, the large rectangle is proportional to the smaller rectangle formed by side b and side a—this is the golden ratio. Some claim that this proportion influenced many artists and early Greek architecture, while others note the variability of picking points in a painting to have golden rectangles superimposed. It is, however, a way of considering the proportionality of a work.

Fundamental Geometric Forms or Shapes

Geometric forms and shapes are the basis for drawing and painting. For example, Piet Mondrian (1872–1944) explored cubism in his work from black and white lines and blocks of primary colors that divided the plane. Other cubists, such as Pablo Picasso, broke with the Renaissance use of perspective to provide an alternative conception of form. Cubists made it possible for the viewer to see multiple points of view simultaneously. Paul Cézanne ignored perspective in some of his work to construct color on the two-dimensional surface. Pointillism was used by Georges Seurat (1859–1891) to create *Sunday Afternoon on the Island of La Grande Jatte*. In pointillism, a series of small, distinct points of color are used to create a painting that relies on the viewer's eye to blend them into a cohesive form. The brain uses the dots to create a solid space. The primary colors are used to create secondary colors for shading and create the impression of a rich palate of secondary colors.

Art deco is characterized by the use of strong geometric forms that are symmetrical. This style of painting was popular in the 1920s and 1930s.

Abstraction and Fractals

Abstraction is an important tenet of mathematics. In mathematical abstraction, the underlying essence of a mathematical concept is removed from dependence on any specific, real-world object and generalized so that it has wider applications. In abstract expressionism, the artist is expressing purely through color and form, with no explicit representation intended. However, that does not mean that abstract art is entirely unstructured. Fractals are one tool used to quantitatively analyze and explain what makes some paintings more pleasing than others. The argument is that, even in an apparently random abstract work, there is an underlying logic or structure that the human brain recognizes as fractal patterns and that it inherently prefers over other works that do not have these patterns. This preference is perhaps because such works are more reflective of the geometry of naturally occurring spaces. For example, physicists Richard Taylor, Adam Micolich, and David Jonas analyzed Jackson Pollock's paintings and found two different fractal dimensions in his work that are mathematically and structurally similar to naturally occurring phenomena, like snow-covered vegetation and forest canopies. In addition to the application of fractals, mathematical concepts like open and closed sets have been used to compare and contrast the work of abstract expressionist artists like Pollock and Wassily Kandinsky to artists like Joseph Turner and Vincent van Gogh, whose works are among those credited with inspiring the expressionist movement.

Mathematical Analysis to Determine Authenticity

Sometimes, the painter of a particular artwork is unknown or disputed, which affects the study of art and the monetary valuation of paintings. Hany Farid and his team created a computer program that uses wavelets to analyze digital images of paintings and map the stroke patterns—some too small to be seen with the naked eye—that characterize an artist's unique style. In one case, known drawings by Pieter Bruegel the Elder were compared to five drawings originally attributed to him. The analysis determined that the five drawings were different from the original eight and also from each other, suggesting multiple creators. Chinese ink paintings are an example in which brush strokes are critical to identification, since they do not have colors or tones to distinguish style. One successful method, tested on the work of some of China's most renowned artists, used a mixture of stochastic models. In another case, fractal geometry was used to question the authenticity of some newly discovered Pollock works, based on his earlier patterns. Radioactive scans and X-ray analysis help to authenticate works by well-known and highly valued masters, such as Johannes Vermeer.

Additional Parallels in Painting and Mathematics

There are many natural parallels in the work of painters and mathematicians. In the same way that painters of different traditions and schools may represent the same scene in drastically different ways, mathematicians may approach the same problem from a variety of disciplines or perspectives. There are also varying degrees of connection to reality in both mathematics and painting. Applied mathematicians and realist painters may be primarily concerned with detailed and faithful representations of the real world in their work, while abstract painters and theoretical mathematicians often work in ways that are logically coherent and consistent, but that do not immediately or obviously connect to the real world. As with art, there is also subjective appreciation of the beauty of mathematics and arguments over what is or is not mathematically valid. Artist Michael Schultheis reported that he was often inspired by mathematical and scientific writing on whiteboards from his days as an engineer, and said, "I constantly revise equations with the Japanese calligraphy brush, rubbing out an area and thus creating a window into the equations. I draw and re-draw new ideas. All of these ideas are analytical. But they also live in the realm of beauty."

Further Reading

Field, J. V. *Piero della Francesca: A Mathematician's Art*. New Haven, CT: Yale University Press, 2005.

Jensen, Henrik. "Mathematics and Painting." *Interdisciplinary Science Reviews* 27, no. 1 (2002).

Robbin, Tony. *Shadows of Reality: The Fourth Dimension in Relativity, Cubism, and Modern Thought*. New Haven, CT: Yale University Press, 2006.

Taft, W. Stanley, and James Mayer. *The Science of Paintings*. New York: Springer, 2000.

Talasek, J. D. "Curator's Essay—Blending the Languages of Mathematics and Painting: The Work of Michael Schultheis." National Academy of Sciences. http://www.michaelschultheis.com/publications/talasek_essay.pdf.

Linda Hutchison

Percussion Instruments

Category: Arts, Music, and Entertainment.
Fields of Study: Geometry; Number and Operations; Representations.

Summary: The vibrations that emanate from percussion instruments vary mathematically based on the type of instrument.

Percussion instruments are characterized by vibrations initiated by striking a tube, rod, membrane, bell, or similar object. Percussion instruments are almost certainly the oldest form of musical instrument in human history. The archeological record of percussion instruments, in particular the *bianzhong* bells of ancient China, give clues to the history of music theory. From a mathematical point of view, percussion instruments are of special interest because—unlike other types of instruments, such as string and wind instruments—the resonant overtones typically do not follow the harmonic series. In the last half of the twentieth century, a question of great interest in applied mathematics has been the famous inverse problem: can one hear the shape of a drum?

Rods and Bars

Some percussion instruments produce a distinct pitch by the vibration of a rod or bar. Examples included the tuning fork (a U-shaped metal rod suspended at its center), a music box (a metal bar suspended at one end), and the melodic percussion instruments such as the xylophone and marimba (suspended at two non-vibrating points or "nodes" along the length of metal or wooden bars). Like vibrating strings, the frequency of the bar's vibration and the pitch of the musical sound it produces are determined by its physical dimensions. In contrast to the string in which the frequency varies inversely with the length, the vibrating bar has a frequency that varies with the square of the length. The resonant overtone frequencies f_n of the vibrating bar are related to the fundamental frequency f_1 by the formula

$$f_n = \alpha \left[n + \frac{1}{2}\right]^2 f_1$$

where the constant α is determined by the shape and material of the bar. In contrast with the harmonic overtone series of vibrating strings, $f_n = n(f_1)$, these inharmonic overtones give percussion instruments their distinct metallic timbre. The overtones of vibrating bars decay at different rates, with rapid dissipation of the higher overtones responsible for the sharp, metallic attack, while the lower overtones persist longer. The bars of the marimba are often thinned at the center, effectively lowering the pitch of the certain overtones, in accord with the harmonic series.

Bells

Like the vibrating bar instruments, the classic church bell possesses highly non-harmonic overtones. These are typically tuned by thinning the walls of the bell along the circumference at certain heights. A distinctive feature in the sound comes from the fact that apart from the fundamental pitch, the predominant overtone of the church bell sounds as the minor third above the prevailing tone. This feature accounts for the somber nature of the sound.

The *bianzhong* bells of ancient China were constructed in a manner that produced two pitches for each bell, depending on the location at which it was struck. In the 1970s, a set of 65 such bells were discovered during the excavation of the tomb of Marquis Yi in the Hubei Provence. The inscriptions on the bells make it clear that octave equivalence and scale theory were known in China as early as 460 B.C.E.

Membranes

Drums are perhaps the most common percussion instrument. Consisting of vibrating membranes (called the "drum heads") stretched over one or both ends of a circular cylinder, drums exhibit a unique mode of vibration, which accounts for their characteristic sound. Mathematical models of vibrating drumheads provide a fascinating application of partial differential equations. The inharmonic overtone frequencies are distributed more densely than for vibrating strings or rods. Further, each overtone is associated with a particular vibration pattern of the drum head. These regions can be characterized by the non-vibrating curves (called "nodes") that arrange themselves in concentric circles and diameters of the drum head.

An important question in the study of spectral geometry asks: "Can one hear the shape of a drum?" In other words, can mathematical techniques be used to work backwards from the overtone frequencies to determine the shape of the drumhead that caused the vibration? The answer, as it turns out, is "not always."

Further Reading

Cipra, Barry. "You Can't Always Hear the Shape of a Drum." In *What's Happening in the Mathematical*

Sciences. Vol. 1. New Haven, CT: American Mathematical Society, 1993.

Jing, M. "A Theoretical Study of the Vibration and Acoustics of Ancient Chinese Bells." *Journal of the Acoustical Society of Americ*a 114, no. 3 (2003).

Rossing, Thomas, D. *The Science of Percussion Instruments*. Singapore: World Scientific Publications, 2000.

Sundberg, Johan. *The Science of Musical Sounds*. San Diego, CA: Academic Press, 1991.

Eric Barth

Popular Music

Category: Arts, Music, and Entertainment.
Fields of Study: Algebra; Measurement; Number and Operations; Representations.
Summary: Popular music can be analyzed and enhanced by mathematical techniques and to some degree the popularity of music can be predicted mathematically.

The interaction between mathematics and popular music goes far beyond the popularity of numbers in song titles, like Tennessee Ernie Ford's "16 Tons" or 2gether's "U + Me = Us (Calculus)." Mathematics is fundamental to musical theory and composition. The twentieth-century subgenres math rock and mathcore are perhaps the most explicitly mathematical compositions, but there are also songs about mathematics concepts. These are usually intended to be humorous or educational, such as "That's Mathematics" by mathematician and musician Thomas Lehrer. Mathematics is also increasingly important to recording and analyzing popular music, including its potential effects on learning. Experimental electronic artist Jamal Moss, founder of record label Mathematics, notes: "Mathematics is the body of sound knowledge centered on such concepts as quantity, music, structure, space, and change—and also the academic discipline that studies them."

Popular Artists

Mathematics in popular music reflects society's often polarized opinions on mathematics. For example, Jimmy Buffet's song "Math Suks" expressed the singer's feelings about the difficulty of mathematical concepts like fractions, algebra, and geometry. Other singers and groups embrace mathematics, like the Texas indie rock band named "I Love Math." Mathematics is often found in album cover art. British band Coldplay's 2005 *X & Y* album featured a cover with colored blocks that spell out "X and Y" in the binary code developed in 1870 by Emile Baudot for use with telegraph systems. Coldplay's lead guitarist Jonny Buckland studied astronomy and mathematics at University College London. Some artists have been criticized for incorrectly using mathematics. Pink Floyd's very popular 1973 album *Dark Side of the Moon* features cover art showing a prism and spectrum. It is correct in depicting some facts, like violet light refracting the most and red the least, but some other aspects are not accurate, such as the relative dispersion of the different colors. Mariah Carey's 2009 album $E = MC^2$, borrowed from Albert Einstein's well-known theory of relativity.

Mathematical Subgenres of Popular Music

Avant-garde composer Iannis Xenakis and post-rock subgenres math rock and mathcore are prominent examples of popular music that relies heavily on mathematics. Xenakis was one of the most significant avant-garde composers of the twentieth century and a grandfather of modern electronic music. His work incorporated mathematical models, such as probability theory, stochastic processes, group theory, set theory, game theory, and Markov chains. He developed algorithms to produce computer-generated music using probability theory and stochastic functions in the 1960s. In his 1966 cello solo "Nomos Alpha," he divided the 24 sections of the piece into two layers. The first layer, consisting of every section not divisible by four, is determined by the 24 orientation-preserving elements of the octahedral group, while the second layer is a more traditional structure. The work has been compared to a musical kaleidoscope, and its structure likened to a fractal.

In the 1990s, post-rock like Slint's *Spiderland* became a dominant genre in experimental rock. Critic Simon Reynolds coined the term "math rock" to describe music that "uses rock instrumentation for non-rock purposes, using guitars as facilitators of timbre and textures rather than riffs and power chords." Math rock bands began to explore the use of dramatically alternating dynamic shifts and unusual time signatures and dissonance, and songs tend to avoid the verse-chorus-verse structure of pop songs. Mathcore developed largely independently of math rock, growing out of hardcore punk and extreme metal, with a huge debt to hardcore pioneers Black Flag.

Mathematics Songs

As of 2010, the Web site M A S S I V E: Math And Science Song Information, Viewable Everywhere is part of the National Science Digital Library and contains over 2,800 mathematical and scientific songs. Popular YouTube songs include mathematical raps and parodies, like "I Will Derive." Hard 'n Phirm's song "Π" rose in popularity because of the 2005 music video by award winning director Keith Schofield. Some songs help students learn mathematics concepts, like multiplication. Other songs showcase the mathematicians who love to sing. The Klein Four Group is a Northwestern University a cappella group who sing about undergraduate and graduate level mathematics. They are most known for their song "Finite Simple Group (of Order Two)."

Self-proclaimed "mathemusician" Lawrence Lesser writes educational songs in order to increase mathematics awareness. Educators often incorporate mathematics songs into their classrooms to enhance student learning of specific concepts and many students use music of various kinds to help them focus while they study mathematical concepts, but these effects are not yet definitively supported or refuted. One study that investigated using jingles to teach statistics concepts found that students who sung several jingles versus reading aloud definitions for the same concepts performed better as a group on a follow-up test. On the other hand, a study that compared classical, popular, and no music to enhance learning found that the students in the three groups performed no differently on a mathematics placement test. This matched findings regarding the effect of music on other academic areas.

Audio Processing

While music production techniques have always allowed a certain amount of alteration and error correction by adjusting the relative levels and balance of the recorded elements, twenty-first century software capabilities have progressed to the point where lower-quality vocals can be processed to professional-sounding quality.

The software package most associated with this is Auto-Tune, released in 1997, and developed by Exxon engineer Harold "Dr. Andy" Hildebrand, who applied seismic data interpretation methods to the analysis and modification of musical pitch. Auto-Tune is an enhancement of existing phase vocoder technology, which uses short-time Fourier transforms, named after mathematician Jean Fourier, to convert time domain representations of sound into time-frequency representations that can be modified before being converted back. Extreme changes can leave tell-tale artifacts in recordings, in the form of a warble like a degenerating audiocassette tape. Audio processing has become standard in many pop albums and on television shows, such as *Glee*. Some well-established singers regularly use Auto-Tune for both albums and in live performances. Other musicians have refused to do so out of fear that it will change the sound enough to make them unrecognizable.

Predicting Popular Song Success

In 2010, Platinum Blue and Music Intelligence Solutions specialize in mathematically predicting hit songs, while services like iTunes and Music IP create suggested playlists or make recommendations. Platinum Blue CEO Mike McCready explained that he and others discovered mathematical patterns in hit songs while trying to build an automated recommendation platform. The algorithm his company uses is based on roughly 30 song traits that are quantified mathematically, such as melody, harmony, beat, tempo, and rhythm. These traits are analyzed for patterns, resulting in groups of songs that are ranked according to probability of success. Hit songs tend to have identifiable similarities, but falling into a particular category is not a guarantee of success. For example, lyrics are an influential song component that are not reliably quantifiable, and aggressive marketing can have an effect not captured by the algorithm. McCready noted: "We figured out that having these optimal mathematical patterns seemed to be a necessary, but not sufficient, condition for having a hit song."

Further Reading

Crowther, Greg, and Wendy Silk. "M A S S I V E: Math And Science Song Information, Viewable Everywhere." *National Science Digital Library*. http://www.science groove.org/MASSIVE/.

Lesser, Lawrence. "Sum of Songs: Making Mathematics Less Monotone!" *Mathematics Teacher* 93, no. 5 (2000).

VanVoorhis, Carmen. "Stat Jingles: To Sing or Not to Sing." *Teaching of Psychology* 29, no. 3 (2002).

Waldman, Harry. "Tom Lehrer: Mathematician and Musician." *Math Horizons* 4 (April 1997).

Xenakis, Iannis. *Formalized Music: Thought and Mathematics in Composition*. 2nd ed. Hillsdale, NY: Pendragon Press, 2001.

Bill Kte'pi

Predicting Divorce

Category: Friendship, Romance, and Religion.
Fields of Study: Algebra; Communication: Data Analysis and Probability.
Summary: Statistical data analysis and mathematical models can be used to predict the likelihood of divorce.

There is a common misconception that one out of every two marriages ends in divorce. The 50% number comes from dividing the number of divorces in a given year (about 1.3 million) by the number of marriages in that same year (about 2.6 million). The mistake is failing to realize that, in any given year, the people getting divorced are probably not the same as those getting married, because the average length of a marriage before a divorce is about eight years (the overall length of marriage, on average, is about 24 years). Hence, those getting married in any given year have an eight-year lag in their projections for divorce. This lag means that the numerator and denominator of the above ratio are not comparable. Instead, experts suggest that about two out of every five marriages end in divorce (or about 40%).

Because of the propensity for some to remain married, for some to divorce more than once, and for some to never marry, only about one out of every five people are predicted to experience a divorce in their lifetime. However, these figures mask the distribution of divorce rates by category—40% of all first marriages end in divorce, 60% of second marriages end in divorce, and 73% of all third marriages end in divorce. There are also some differences by age group, with divorce rates highest for those in their early 20s and declines steadily in subsequent age groups.

There are two main ways to predict divorce: empirical (or statistical) methods that take advantage of data gathered on married and divorced couples; and mathematical models that try to make a priori predictions of future divorce using features of existing marriages or theoretical assumptions based on extensive work in the area.

Empirical Methodology

Empirical work suggests that indicators predicting divorce can be separated into two groups: factors present before marriage and factors that occur within the marriage. Some of the more common risk factors brought into a marriage include parental history of divorce (children of divorced parents are more likely to divorce), educational attainment (those with lower levels of education are more likely to divorce), and age (those who marry younger are more likely to get divorced). The risk factors that arise within the marriage include communication styles (couples with poor or destructive communication have a greater chance of divorce), finances (couples with financial problems, including a large disparity in spending habits, disposable income, and wealth goals, are at a greater risk for divorce), infidelity, commitment to the marriage (a lack of commitment or a dissimilarity in the amount of commitment often leads to divorce), and dramatic change in life events.

Mathematical Models

Mathematical models seek to discover features of current relationships that will put a couple at risk for future divorce. Professor John Gottman argues that the way couples communicate can often predict divorce. His research, which is based on analyzing hundreds of videotaped conversations between married couples, claims a 94% accuracy rate. The work also monitors pulse rates and other physiological data that, when combined with the observations, leads to what he calls the "bitterness rating." The rating is based on six signs. The first sign posits that when a conversation starts with accusations, criticisms, or negativity, the discussion is likely to end badly. However, he argues that the opposite is also true. The second sign encompasses four patterns of negative interaction that can be deleterious to a marriage: criticism, contempt, defensiveness, and stonewalling. The third sign is "flooding," in which negativity of one partner overwhelms the positive feelings of the spouse until there is virtually nothing left but discontent. The fourth sign recognizes that physiological changes, such as increases in adrenaline and blood pressure, often lead to feelings of entrapment and serve to poison an otherwise benign conversation. The fifth sign identifies the fact that some marital discord is unchanged by the repeated attempt by one partner to repair the damage done to the relationship. Finally, the sixth sign involves one or both people rewriting the history of their relationship to be largely negative. Once people reach the sixth sign, Gottman argues, divorce is likely.

Further Reading
Booth, Alan, and John N. Edwards. "Age at Marriage and Marital Instability." *Journal of Marriage and the Family* 47 (1985).
Gottman, John, and Nan Silver. *The Seven Principles for Making Marriage Work*. London: Orion, 2004.
Martin, Teresa Castro, and Larry L. Bumpass. "Recent Trends in Marital Disruption." *Demography* 26 (1989).
South, Scott, and Glenna Spitze. "Determinants of Divorce Over the Marital Life Course." *American Sociological Review* 51 (1986).
Wolfinger, Nicholas H. "Trends in the Intergenerational Transmission of Divorce." *Demography* 36 (1999).

Casey Borch

Puzzles

Category: Games, Sport, and Recreation.
Fields of Study: Algebra; Geometry; Number and Operations.
Summary: Because problem solving is a core activity of mathematics, it lends itself well to puzzles.

A puzzle is a question, problem, or contrivance designed to challenge and expand the mind and perhaps test ingenuity. Puzzles have been found in virtually all cultures and all historic periods, even in mythology. According to legend, the Sphinx prevented anyone from entering Thebes who failed to find the correct answer to the question: What is it that has four feet in the morning, two at noon, and three at twilight?

Mathematicians have long created puzzles and explored their solutions for research and applications. They have also created puzzles for purely recreational purposes. Teachers in many subjects within and outside mathematics use puzzles in the classroom.

There are a number of ways in which words and arrangements of letters or objects are used to create puzzles. Some problems in the Rhind Mathematical Papyrus (1650 B.C.E.) are seen as puzzles. One example is a rhyme that also appears in Leonardo Pisano Fibonacci's 1202 work *Liber Abaci* and is still popular today. Here is a modern version:

> As I was going to St. Ives,
> I met a man with seven wives.
> Each wife had seven sacks,
> Each sack had seven cats,
> Each cat had seven kits.
> Kits, cats, sacks, wives,
> How many were going to St. Ives?

One may only assume that the narrator was going to St. Ives, not necessarily the other travellers. Mathematically, logic, branching diagrams, multiplication, and addition can be used to determine the final solution.

Traditional in several cultures, namely in Africa, is the Crossing Problem. The following is a version from Alcuin of York (735–804):

> A man wishes to ferry a wolf, a goat, and a cabbage across a river in a boat that can carry only the man and one of the others at a time. He cannot leave the goat alone with the wolf nor leave the goat alone with the cabbage on either bank. How will he safely manage to carry all of them across the river?

To solve this problem, one must recognize that the man may carry an item back and forth across the river as many times as needed and ultimately find appropriate combinations and sequencing. Dynamic versions of this game appear online, adding visual and tactile components to the solving process. Extensions of this problem include adding more items to the list, increasing the size of the boat to carry more items, and adding an island in the middle of the river where objects may be placed. Mathematicians such as Luca Pacioli, Niccolo Tartaglia, Claude-Gaspar Bachet, and Edouard

Lucas investigated this problem. A well-known medieval task consisted of arranging men in a circle so that when every *k*-th man is removed, the remainder shall be a certain specified man. Several authors commented on this, from Girolano Cardano in the sixteenth century to Donald Coxeter in the twentieth century.

Word Puzzles

Anagrams have a long and mysterious history, being seen as source of ludic pleasure but are also believed by some to possess mystic powers. Inside a word or phrase, another one is hiding that one can get by permuting the letters in a different order. For instance, the letters in the word "schoolmaster" may be rearranged to form the related phrase "the classroom."

Lewis Carroll (1832–1898) invented a forerunner of the crossword: the "doublet." There are two words presented to the solver, who is required to change one word to the other by replacing only one letter at a time, forming a legitimate word with each transformation. One of his examples is to change "HEAD" into "TAIL," which can be done via the following sequence: "HEAL," "TEAL," "TELL," "TALL," and "TAIL."

Visual Puzzles

Visual puzzles are also popular, such as optical illusions, which have long been investigated by mathematicians. Some of these address mathematical questions in disciplines like geometry and visualization, including figures that appear to be impossible.

Figure 1. Are the two dark lines parallel?

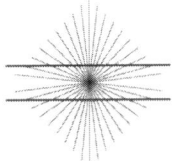

Figure 2. An illustration of an "impossible" object.

Samuel Loyd (1841–1911) is referred to by some as "America's greatest puzzlist." He reputedly created thousands of puzzles. Some of his inventions were very original, like the Get Off the Earth puzzle. There are 13 men in the figure on the left. Rotating the puzzle, as shown in the figure on the right, produces a drawing that has 12 men. What happened to the 13th man?

Figure 3. The Get Off The Earth puzzle.

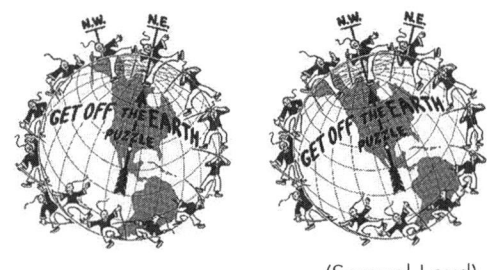

(Samuel Loyd)

Arithmetic Puzzles

Numerical relations and arithmetical principles are often found in puzzles. "Magic squares," which are square arrays of consecutive numbers with constant sum in columns, rows, and diagonals, illustrate this clearly. One of the oldest, the Chinese *lo-shu*, dates back thousands of years. Leonhard Euler's (1707–1783) work on Latin Squares, which are arrays of symbols with no repetitions in rows or columns, is one of the foundations of *Sudoku* puzzles, which appeared in a U.S. magazine in the 1970s but became famous first in Japan and then in the world. Tartaglia (1500–1557) presented the following numerical problem: A dying man leaves 17 horses to be divided among his three sons in the proportion 1/2 : 1/3 : 1/9. Can the brothers carry out their father's will? Since 17 is not a multiple of 2, 3, or 9, there is no solution that would give all of the sons a whole number of horses.

Some authors shared problems, even if they lived in different centuries. Fibonacci (1170–1250), Tartaglia, and Bachet (1581–1638) all investigated the question:

> If you have a balance, what is the least number of weights necessary to weigh any integer number of pounds from 1 to 40? (Assume you can put weights in either side of the balance.)

"Cryptarithms," created for training the calculating mind in 1913, were very popular in the twentieth century.

In a cryptarithm, one is asked to find the digits erased from a valid calculation. Later, prolific English puzzle inventor, Henry Dudeney (1857–1930), substituted letters for the unknown numbers to create another layer of meaning. In his first example of an "alphametic" is the equation: SEND + MORE = MONEY, where each letter represents a different digit, and the addition is correct.

Rearrangement Puzzles

Some dissection and rearrangement puzzles are based on mathematical principles. Archimedes of Syracuse (287–212 B.C.E.) may have created a 14-piece puzzle, the "Stomachion," as part of his research. It resembles a version of a "Tangram," a Chinese puzzle that became very popular in the nineteenth century in the West and is often used in mathematics classrooms in the twenty-first century to investigate dissections and concepts like the Pythagorean Theorem, named for Pythagoras of Samos. The Fibonacci sequence relation

$$(F_n)^2 = F_{n-1}F_{n+1} + (-1)^{n-1}$$

with $n = 6$ can be used to create a dissection puzzle. Larger values of n generate similar, more impressive puzzles, where the difference of area between a large square and a large rectangle is always included. Some dissection puzzles may lead to optical illusions when the pieces do not fit exactly together, leading to two figures composed of the same pieces that have different areas.

Figure 4. An 8-by-8 square and 5-by-13 rectangle made with the same pieces?

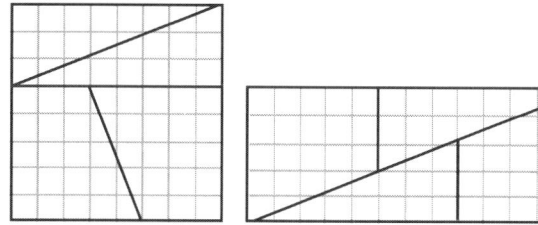

Topological Puzzles

Ring and string puzzles as well as knotted puzzles are examples of topological puzzles, where no discontinuous deformations like cutting the string are allowed. In his *De Viribus Quantitatis* (c. 1500), cited as the oldest book in recreational mathematics, Luca Pacioli (1445–1517) describes the Chinese Rings, a topological puzzle still popular in the twenty-first century.

Figure 5: A modern version of the Chinese Rings puzzle

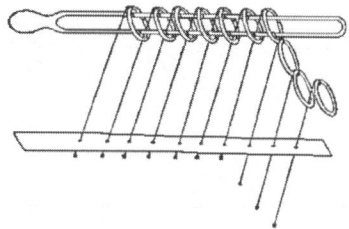

Euler's name is linked to several puzzles. He solved the Bridges of Konigsberg Problem, and this work of his is usually seen as the starting point of topology and graph theory.

Figure 6: The Bridges of Konigsberg Problem: is it possible to cross all the bridges only once?

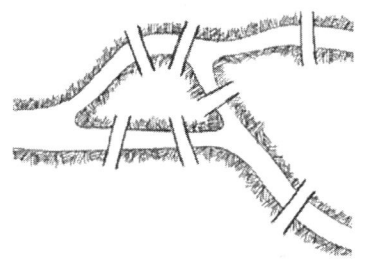

The concept of the Eulerian graph is rooted in Euler's resolution of the Bridges of Konigsberg problem.

Movement Puzzles

Numerous puzzles involve patterned movement within some type of framework, and solutions sometimes involve mathematical techniques like numbering, recursion, group theory, and determinants. Peg Solitaire traces its origins from seventeenth-century France. It is a game where a board has all its holes occupied with pegs except for the central one. The objective is, making valid moves (small jump capture), to empty the entire board but for a solitary peg in the central hole (see Figure 7).

The Towers of Hanoi is a puzzle invented in 1883 by N. Claus, a pseudonym of the mathematician Edouard Lucas (1842–1891). A pile of discs of decreasing radius lays on one of three poles. Moving one disc at a time, without letting a bigger disc rest on a smaller one, the solver is asked to change the pile from one pole to another (see Figure 8).

Figure 7. Peg Solitaire: starting and target position.

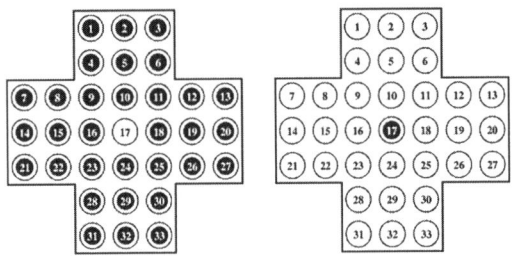

Figure 8. Towers of Hanoi: starting and target positions.

The recursive character of the solution to this puzzle makes it somewhat similar to the Chinese Rings.

Other Puzzles

The chessboard is a rich source of puzzles that attracted many mathematicians. In the Knight Tour problem, a knight must visit all the squares of the board just once. Euler is one mathematician who published a solution. Mathematician Johann Carl Friedrich Gauss (1777–1855) was attracted by the 8-Queen Problem, in which eight queens must be placed on a chessboard so they cannot capture any other queen. Some mathematicians have used determinants to solve this problem.

Figure 9. The 8-Queen Problem: one solution.

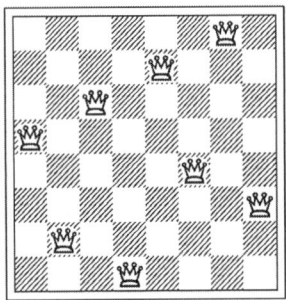

The nineteenth century produced a popular puzzle named "15." It consists of a sliding device, a 4-by-4 array with the numbers one through 15 and an empty cell. The puzzle was scrambled and the solver was required to transform the scrambled order back to the natural order with the empty cell in the last position. Sam Loyd offered $1,000 to whoever could reorder a scrambled 14 and 15 in an otherwise solved puzzle. The prize was never claimed. The impossibility of this challenge can be understood when phrased in the language of group theory.

Figure 10: The "impossible" task.

1	2	3	4
5	6	7	8
9	10	11	12
13	15	14	

Another very mathematical puzzle that captivated the world was Rubik's Cube, created by Hungarian architect Erno Rubik in the 1970s that became the best selling puzzle in history. A 3-by-3-by-3 cube, with differently colored faces, moves by slices, getting scrambled with just a few moves. To find the way back to the starting position is an incredible challenge. This toy puzzle is used to illustrate many group theory concepts. On the other hand, knowledge of group theory facilitates the understanding of the puzzle itself.

Since ancient times, descriptions of "mazes" that must be traversed in a particular pattern of moves have abounded in legend and literature. The Minotaur–Theseus tale is one such example. Stone and hedge labyrinths may still be found in places like Europe and many puzzle books contain paper mazes. Some mazes can be understood using what is known as "level sequences."

The "jigsaw puzzle" was invented in England in the mid-1870s as a pedagogical device. Children were asked to rebuild maps. In the twentieth and twenty-first centuries, jigsaw puzzles expanded to include three-dimensional jigsaw puzzles, including spherical three-dimensional puzzles, and two-dimensional jigsaw puzzles that are all one color that have all the pieces cut to the same shape. This last style of puzzle is related to tiling. Another mathematical question is how to optimally and efficiently design and cut out puzzle pieces according to certain specifications.

Puzzle designer Scott Kim is considered by some to be a master of symmetry. He has diverse interests in many fields, including mathematics, computer science, puzzles, and education. When discussing these interests, he emphasizes the ties between them rather than their differences. One of his creations is an ambigram to honor of the great Martin Gardner (1914–2010), who invented many puzzles and is known for his recreational mathematics works. An ambigram is a figure that appears the same when rotated 180 degrees or viewed upsidedown.

Further Reading

Danesi, Marcel. *The Puzzle Instinct: The Meaning of Puzzles in Human Life.* Bloomington: Indiana University Press, 2002.

Dedopulos, Tim. *The Greatest Puzzles Ever Solved.* London: Carlton Books, 2009.

Olivastro, Dominic. *Ancient Puzzles: Classic Brainteasers and Other Timeless Mathematical Games of the Last 10 Centuries.* New York: Bantam Books, 1993.

Petkovic, Miodrag. *Famous Puzzles of Great Mathematicians.* Providence, RI: American Mathematical Society, 2009.

Sam Loyd's Puzzles. http://www.samuelloyd.com/gallery.html.

Scott Kim Puzzlemaster. "Inversions Gallery." http://www.scottkim.com/inversions/gallery/gardner.html.

Slocum, Jerry, and Jack Botermans. *New Book of Puzzles: 101 Classic and Modern Puzzles to Make and Solve.* New York: W.H. Freeman, 1992.

———. *The Tangram Book: The Story of the Chinese Puzzle With Over 2,000 Puzzles to Solve.* New York: Sterling Pub. 2003.

Slocum, Jerry, and Dic Sonneveld. *The 15 Puzzle: How It Drove the World Crazy; The Puzzle That Started the Craze of 1880; How America's Greatest Puzzle Designer, Sam Loyd, Fooled Everyone for 115 Years.* Beverly Hills, CA: Slocum Puzzle Foundation, 2006.

Spencer, Gwen. "A Conversation with Scott Kim." *Math Horizons* 12 (November 2004).

JORGE NUNO SILVA

Pythagorean and Fibonacci Tuning

Category: Arts, Music, and Entertainment.
Fields of Study: Algebra; Measurement; Representations.
Summary: The relationship between mathematics and music led to several tuning systems.

A musical scale is a sequence of ordered notes used to construct music compositions. Scales can be classified according to their starting point, the intervals between their notes, or the number of notes they contain. Instruments may be tuned according to many possible systems. There are close mathematical connections between musical scales, tuning systems, and number theory, as well as dynamical systems. Mathematics also plays a critical role in designing playable and efficient keyboards for instruments that will be tuned to something other than the standard eight-note Western scale.

Most Western music uses an eight-note "octave" scale (do, re, mi, fa, sol, la, ti, do), where the two "do" notes have the same tone but different pitches. The piano keyboard is set up in the C major key, where the white keys starting with C correspond to the eight notes in the octave.

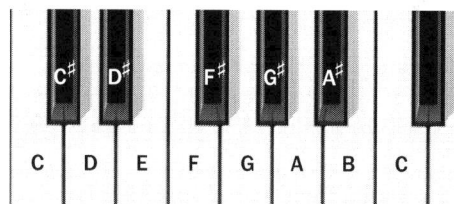

There are also tones between some of the notes on the scale, represented on the piano by the black keys. Counting from C to B, there are 12 equal semitones in the chromatic scale of Western music.

To tune an instrument with strings, the lengths of the strings are adjusted to produce the correct pitch. Pythagoras of Samos (570–495 B.C.E.) is credited with realizing two things that allowed him to calculate the string lengths for the 12 semitones of the chromatic scale:

1. A string that is half as long produces the tone that is one octave higher. A string that is twice as long produces a tone that is one octave lower.

Figure 1.

F♯	G	G♯	A	A♯	B	C	C♯	D	D♯	E	F	F♯	G	G♯	A	A♯	B	C
11	3	14	6	17	9	1	12	4	15	7	18	10	2	13	5	16	8	19

2. A string that is two-thirds as long produces a tone that is up five notes (called a *fifth*, or *do-sol* interval), seven semitones higher in the 12-tone chromatic scale.

Pythagoras saw that seven and 12 share no common factors and that he could use this fact to generate the lengths of all 12 strings in the chromatic scale.

1. Start with a string that sounds like a C note.
2. Cut a string that is two-thirds of the C string to give G.
3. Cut a string that is twice as long as G, yielding the same tone down an octave.
4. Cut a string two-thirds of this new lower G to give D.
5. Cut a string two-thirds as long as D to give A.
6. Cut a string twice as long as A, yielding A down an octave.
7. Cut a string two-thirds of the lower A to give E.
8. Cut a string two-thirds of E to give B.
9. Cut a string twice as long as B, yielding B down an octave.

Continue in this pattern, shortening a string to two-thirds to produce new higher notes and doubling the string when needed to avoid going past the top of the octave. After 19 steps, all of the strings of the C to C octave are determined, as well as a few extra notes below C (see Figure 1).

Called the "circle of fifths," this method of tuning by shortening the string to move up seven semitones (and back 12 when needed) would not work if the two numbers involved shared a common factor, such as four and 12. Not all of the semitones would be "hit" in that case.

Equal Tuning

Pythagoras was a little off when he assumed that a string two-thirds as long would produce the seventh semitone. In actuality, using irrational numbers (something Pythagoras did not believe in), the lengths of string needed to produce all of the semitones can be found more precisely. Starting with a string of length two, one can factor two into 12 equal parts or "twelfth roots." This method of tuning, used in the twenty-first century for most music, is called "equal tuning" (see Figure 2). The values of these irrational numbers to three decimal places show that the fifth note (or seventh semitone) string, G, is actually slightly more than two-thirds of the C string: two-thirds of a string of length 2 would yield a G string of length 1.333 rather than the equal tuning length of approximately 1.335. This little bit of difference is magnified when the circle of fifths technique is used to tune the strings, yielding notes that sound flat.

Other Tuning Systems

Between Pythagoras's time and the twenty-first century, a number of other tuning strategies were developed as music and mathematics knowledge grew. Popular in the medieval age, for example, was "just" tuning, which differs from both Pythagorean and equal tuning. To use equal tuning in the twenty-first century, one does not have to physically measure strings precisely; equipment can be used to measure the fundamental frequency

Figure 2.

C	C♯	D	D♯	E	F	F♯	G	G♯	A	A♯	B	C
$\left(\sqrt[12]{2}\right)^{12}$	$\left(\sqrt[12]{2}\right)^{11}$	$\left(\sqrt[12]{2}\right)^{10}$	$\left(\sqrt[12]{2}\right)^{9}$	$\left(\sqrt[12]{2}\right)^{8}$	$\left(\sqrt[12]{2}\right)^{7}$	$\left(\sqrt[12]{2}\right)^{6}$	$\left(\sqrt[12]{2}\right)^{5}$	$\left(\sqrt[12]{2}\right)^{4}$	$\left(\sqrt[12]{2}\right)^{3}$	$\left(\sqrt[12]{2}\right)^{2}$	$\sqrt[12]{2}$	1
2	1.888	1.782	1.682	1.587	1.498	1.414	1.335	1.260	1.189	1.122	1.059	1

(related to the pitch) of the sound wave generated by the string in order to tighten the string to the correct length.

There is also a method of tuning based on the Fibonacci series of Leonardo Pisano Fibonacci, which has been analyzed by English mathematician Sir James Jeans. The numbers in the musical Fibonacci series (2, 5, 7, 12, 19, . . .) can be generated by increasingly long series of musical fourths and fifths from the octave scale. An interval of two tones that are a fifth apart, such as F and C, have a frequency ratio of three-halves. The next fifth is a G, which is musically very close to the original F, but an octave higher, so the two-tone scale is left as F and C. Extending the fifths to a five-tone scale gives F, C, G, D, and A. This would be followed by E, which is again almost the initial F. A slight modification made by slightly raising all the tones (after the initial F) would create a five-note equal tuning scale. Increasingly larger scales can be made by continuing this pattern.

Further Reading
Ashton, Anthony. *Harmonograph: A Visual Guide to the Mathematics of Music.* New York: Walker & Co., 2003.
Hall, Rachel W., and Kresimir Josic. "The Mathematics of Musical Instruments." *American Mathematical Monthly* 108, no. 4 (2001).
Jeans, James. *Science and Music.* New York: Dover Publications, 1968.

Holly Hirst

Quilting

Category: Arts, Music, and Entertainment.
Fields of Study: Geometry; Measurement; Representations.
Summary: Quilting can incorporate and help teach mathematical concepts, such as symmetry and tessellations.

Quilting is a needlework technique in which two layers of fabric are sewn together, usually with an inner layer of padding (called "batting") between them. Often, one or both outer layers are formed by sewing together (or "piecing") smaller pieces of fabric. Sometimes, designs are appliquéd (sewn onto a larger piece of fabric) or embroidered on the quilt. The quilting itself (the stitches holding the layers together) is often also decorative. Many traditional quilt designs display mathematical concepts, such as symmetry and tessellations, that generalize into the abstract mathematics of group theory and tiling theory. In diverse parts of the world, people create quilts not only to warm the body at night, but also to use as clothing, furnishings, or to share family or cultural history. A carving of an ancient Egyptian Pharaoh figure containing what may be a quilt and a quilted carpet found in the mountains of Mongolia dates to approximately the first century. Directions can be found to quilt coded designs that may have been used on the Underground Railroad.

Quilt Designs
Some traditional quilts are "crazy quilts" in which scraps of fabric are sewn together in no particular pattern. Others are formed of similar or identical square "blocks," each of which may be pieced together. Often, quilt patterns involve careful measurement (using common fractions) in the cutting and sewing of the pieces.

Quilt designs are often symmetrical—the entire design can be folded in half along a line such that one half falls directly onto the other half. Each half is a reflection of the other along that line, which is called a "line of symmetry." These lines may be vertical, horizontal, or diagonal. Some quilt blocks, such as the traditional Amish Star, are symmetric along many lines. Quilts and quilt blocks may also have rotational symmetry—the design can be rotated around a point through less than a full rotation in a way that leaves the overall design unchanged. Quilts in the Hawaiian Islands are known for their distinctive radial symmetry.

Mathematics generalizes this everyday concept of symmetry. A mathematical object (not necessarily a geometric shape) is symmetric with respect to a particular mathematical operation if the operation, applied to the object, preserves some property of the object. A mathematical group consists of a set of operations that preserve a given property of a given object. Group theory is central to abstract algebra and has many applications.

Fabric quilts, construction paper versions, or computerized models of quilt designs have been used to introduce students as early as elementary school to geometric concepts, such as symmetry and transformations. They help children develop, at a basic level, fundamental algebraic properties, such as inverse, identity, and equivalence. Students also make quilts

Many traditional quilt designs display mathematical concepts, such as symmetry and tessellations, that generalize into the abstract mathematics of group theory and tiling theory. (iStockphoto)

to explore many other concepts, such as the Pythagorean theorem, polar coordinates, group theory, the Fibonacci sequence, and Pascal's triangle, named after mathematician Blaise Pascal.

Tessellations

A tessellation (or tiling) is an infinitely repeating pattern composed of polygons covering a plane without any openings or overlaps. Many quilt designs are formed from tessellations. A regular tessellation uses one polygon with equal sides and equal angles, such as equilateral triangles, squares, or regular hexagons. For example, the traditional Grandmother's Flower Garden and Honeycomb quilt designs use tessellations of regular hexagons. Many modern watercolor quilts use tessellations of one-inch squares.

A semi-regular tessellation uses a combination of squares, triangles, and hexagons that are arranged identically around each vertex. Demi-regular tessellations, with two vertices in each repetition, form more complicated quilt patterns. Many quilt blocks, such as Log Cabin variations, consist of non-regular tessellations.

Mathematicians have generalized tiling theory to higher dimensional Euclidean spaces and to non-Euclidean geometries. These generalizations reveal links to group theory and to classical problems in number theory. Much of the art of M.C. Escher is based on non-Euclidean tessellations.

Other Designs

Contemporary quilters like mathematician Irena Swanson have also incorporated other mathematical concepts in their designs, such as infinite geometric series and fractals, as well as portraits of mathematicians. Mathematician Gwen Fischer created quaternionic quilts to visually showcase the algebraic structure of the group. For example, the lack of reflection symmetry across the main diagonal highlights the lack of commutativity of the group elements.

Further Reading

Fisher, Gwen. "Quaternions Quilt." *FOCUS* 25, no. 1 (2005).

Meel, David, and Deborah Youse. "No-Sew Mathematical Quilts: Needling Students to Explore Higher Mathematics." *Visual Mathematics* 10, no. 2 (2008).

Paznokas, Lynda. "Teaching Mathematics Through Cultural Quilting." *Teaching Children Mathematics* 9 (2003).

Rosa, Milton, and Daniel Orey. "Symmetrical Freedom Quilts: The Ethnomathematics of Ways of Communication, Liberation, and Art." *Revista Latino Americana de Etnomatemática* 2, no. 2 (2009).

Venters, Diana, and Elain Ellison. *Mathematical Quilts— No Sewing Required*. Emeryville, CA: Key Curriculum Press, 1999.

Bonnie Ellen Blustein

Racquet Games

Category: Games, Sport, and Recreation.
Fields of Study: Algebra; Data Analysis and Probability; Geometry.
Summary: The equipment, game play, and scoring of racquet sports can be analyzed using mathematical concepts, such as vector operations and probability.

Racquet games include sports such as tennis, badminton, squash, and table tennis, as well as other less popular games like real tennis, racquets, and racquetball. Mathematics has many roles to play in these games—from equipment testing and court marking to training and analysis of play.

For example, the scoring system in tennis is not a simple counting or linear progression. Mathematicians model a ball's spin in multiple axes, along with trajectories and deflections, as functions of other variables. Markov chains and vector operations can be used to analyze the progression of games and both probability and statistical methods are used to describe performance, seed players for competition, and predict outcomes of matches.

Racquets

Racquet weight distribution, shape, and string material are important factors in the resultant power, accuracy, and comfort of a racquet. Increasing power, for example, can lead to a decrease in accuracy and it is important to balance these properties. Computer-aided design is the natural choice for this process because of its fast and powerful recalculation abilities.

Projectiles

Racquet sport projectiles such as balls and shuttlecocks are subject to strict regulations and must adhere to these for as long as possible at the highest levels of play. For example, the World Squash Federation allows balls that are 40 millimeters in diameter and each must be tested at 23 degrees Celsius (73 degrees Fahrenheit) and 45 degrees Celsius (113 degrees Fahrenheit), room temperature and play temperature, respectively. There are several dot grades according to level of rebound but an average squash ball rebounds at around 30% (dropped from a height of 3.2 feet, it should reach 12 inches on the bounce). A tennis ball rebounds at around 50%, although changes in ambient air pressure (because of altitude) can affect this figure. Table tennis balls rebound at 85%.

A popular way to gauge the overall performance of these projectiles is to measure their maximum speed. Tennis balls seem to hold the record for the being the fastest, and indeed Andy Roddick can propel a tennis ball very fast (152 miles per hour). However, the fastest badminton stroke left the racquet at over 186 miles per hour. This figure seems counterintuitive because a shuttlecock slows down much more quickly than a tennis ball.

Training

One of the most important roles for mathematics in racquet sports is in training. Sports science researchers study muscle and joint strain and develop nutritional guidelines that allow the player to remain comfortable and energetic during play. Of the racquet sports, squash is regarded as the most intense—players burn roughly 50 percent more calories per hour than badminton or tennis. However, tennis games can run several hours, whereas badminton and squash games are typically decided in under an hour. The total number of calories burned is the product of the calories per hour and the number of hours.

Scoring

In all of the major racquet sports (and many others), a feature of the scoring system may mean that the player who wins more individual points or rallies can still lose the match. Consider the scores of the 1972 British Open final decided by the best of five games, each played to nine points: 0–9, 9–7, 10–8, 6–9, and 9–7. The loser (Geoff Hunt) scored 40 points and won two games; the winner (Jonah Barrington) scored 34 points, won three games and the title.

The same quirk appears in any scoring system where victory is decided by the most wins over a specific number of games. In tennis, this feature exists on two levels. It is possible to win more points and more games but still lose the match. For example, if a match ends 6–4, 0–6, 6–4, 0–6, 6–4, the winner wins 18 games, the loser wins 24 games. The maximum difference in points or rallies in this case is 60 (72–132) in favor of the loser.

Further Reading

Gallian, Joseph. *Mathematics and Sports*. Washington, DC: Mathematical Association of America, 2010.

Havil, Julian. *Nonplussed! Mathematical Proof of Implausible Ideas.* Princeton, NJ: Princeton University Press, 2007.

Lees, A., D. Cabello, and G. Torres, eds. *Science and Racket Sports IV*. New York: Routledge, 2009.

Lees, A., J. F. Kahn, and I. W. Maynard, eds. *Science and Racket Sports III*. New York: Routledge, 2004.

Sadovskii, L. E., and A. L. Sadovskii. *Mathematics and Sports*. Providence, RI: American Mathematical Society, 2003.

Eoin O'Connell

Rankings

Category: Games, Sport, and Recreation.
Fields of Study: Number and Operations; Measurement.
Summary: Ranking is a widely used to create ordered lists of people or objects, and there are many ways to assign and analyze ranks.

Throughout human history, people have been ordering objects into hierarchies based on criteria such as measurements or qualitative properties. In the twenty-first century, people rank many objects, such as quarterbacks, political candidates, and restaurants. Every spring, high school seniors eagerly wait to see who will be the valedictorian, or top-ranked student, of their high school class. However, there is not usually a single unique ranking for a set of objects, since ranks depend on the criteria selected and the specific method in which they are combined. *US News and World Report* aggregates multiple quantitative and qualitative indicators in its annual ranking of colleges. Mathematicians use a variety of techniques to study ranking, such as algebra, geometry, graph theory, game theory, operations research, and numerical methods. An entire subset of statistical techniques based on ranks, called *nonparametric* or *distribution-free tests*, are used to transform and analyze data that do not conform to the assumptions or parametric tests.

These techniques are often used in the social sciences. There are also debates about whether ranks are true numbers, given that the spacing between ranks need not be equal in the manner of most common measurement scales. For example, the difference between one inch and two inches is the same as between two inches and three inches. The difference between first and second place, however, is not necessarily quantitatively or qualitatively the same as the difference between second and third place.

Tiebreakers

Some ranking strategies result in ties between one or more individuals. Sometimes there is a tiebreaker, and other times there is not. The ranking of items occurring after the tie can vary depending on the type of ranking used. The most common is called standard competition ranking, where a gap is left in the numbering after the tie takes place corresponding to the number of elements in the tie. For example, if there were six items and a three-way tie for second occurred, the ranking would be given as "1, 2, 2, 2, 5, 6" with third and fourth place omitted. Some methods, especially those used in statistical analysis, assign an average rank. In a three-way tie for second place out of six objects, the assigned rankings would be "1, 3, 3, 3, 5, 6," since the average of 2, 3, and 4 is 3.

Sports

Athletic competitions are one very visible use of rankings. During the ancient Olympic Games, athletes would compete in events, such as running, boxing, and the pentathlon, to determine which athletes were better than others. Ultimately, they would be ranked by their performance in these events. Even during the modern Olympics, though the events are more numerous and athletes generally compete in only a few events, the result is a ranking of the best athletes, with prizes being awarded to the top three finishers. There are rankings for other sports as well. For example, the Associated Press ranks the top 25 NCAA football teams by polling sportswriters across the nation. Each writer creates a personal, subjective list of the top 25 teams from all eligible teams (more than 25). The individual rankings are then combined to produce the national ranking by giving a team 25 points for a first place vote, 24 points for a second place vote, and so on down to one point for a 25th place vote. Teams are also regularly ranked by their number of wins or other game-related metrics, as are individual players.

Tests

Rankings also occur on standardized tests. Rather than give each individual a unique rank, tests such as the SAT separate the scores into percentages and then rank test takers according to the percentage they fall into. Percentile ranks can also be seen in other places, such as height and weight charts for children. Whereas many rankings place an emphasis on small numbers (it is better to be ranked first or second than twenty-fifth), percentiles are considered in the opposite manner—a larger value percentile ranking is a better rank. Percentiles indicate what percentage of the test-taking group performed the same or worse than a test-taker in that percentile. For example, being in the 57th percentile would indicate that 57 percent of the test takers scored the same or worse. When considering rakings, it is important to determine how the ranking is arranged to properly interpret the data.

Other Mathematical Connections

The word "rank" carries many specific definitions in various fields of mathematics. For example, the rank of a matrix is the number of linearly independent rows or columns. In graph theory, the rank of a graph is the number of vertices minus the number of connected components. Other definitions of rank can be found in set theory and Lie algebra (named for mathematician Sophus Lie). In chess, a game studied by many mathematicians, a rank is a row on the chessboard.

Further Reading

Gupta, Shanti, and S. Panchapakesan. *Multiple Decision Procedures: Theory and Methodology of Selecting and Ranking Populations*. Philadelphia: Society for Industrial Mathematics, 2002.

Marden, John I. *Analyzing and Modeling Rank Data*. New York: Chapman & Hall, 1995.

Winston, Wayne. *Mathletics: How Gamblers, Managers, and Sports Enthusiasts Use Mathematics in Baseball, Basketball, and Football*. Princeton, NJ: Princeton University Press, 2009.

CHAD T. LOWER

Representations in Society

Category: School and Society.
Fields of Study: Connections; Representations.
Summary: Symbols, equations, and images are all used to teach mathematical concepts and to convey mathematical information in society.

Representations are at the forefront of the focus standards of the National Council of Teachers of Mathematics to improve mathematics teaching and learning. Representations allow students to see and experience mathematics from different perspectives. The role of multiple representations in promoting students' conceptual understanding of mathematics has long been emphasized by researchers. Thus, representations are among the essential parts of mathematics lessons. Further, in the twenty-first century, even people who had very little exposure to mathematics in school will encounter various mathematical representations in their daily lives. Familiarity with mathematical representations or representational literacy has become an essential skill. Many mathematical concepts are defined in terms of representations. A function may be represented by a Taylor series of infinite terms, which is named after Brook Taylor. There is also an entire branch of mathematics called "representation theory" that expresses algebraic structures using linear transformations.

Representations

Mathematics has its own native beauty and inspirational aesthetic to represent the physical world and the world of intellect. One of the strengths of mathematics is its resources to seek for new solutions and explore frameworks to answer problems related to the real world. To achieve this goal, mathematical representations in society should be explored and important ideas of modern mathematics should be communicated properly. Representations in mathematics can be described as constructs that symbolize or correspond to real-world mathematical entities, features, or connections. Gerald Goldin broadly defined representations as any configuration of characters, images, or concrete objects that can symbolize or represent something else. Representations take various forms, such as informal representations used in preschool settings or more formal representations used in mathematics classrooms or by mathematicians. For example, children represent groups of five with their hand or, even further, they develop proportional thinking as they relate five fingers to one hand and 10 fingers to two hands. More formally, mathematics students or mathematicians use mathematical equations, for example, to represent curves or relationships among financial variables.

Internal and External Representations

Representations can be both internal and external in nature and can be created by forming individual representations, such as letters, numbers, words, real-life objects, images, or mental configurations. Internal representations are mental images or cognitive constructs of individuals that relate to external representations or to experiences in the external world. James Kaput referred to internal representations as *mental structures* and defined them as instruments that are used to organize and manage the flow of an individual's experience. Internal representation systems exist within the mind of an individual and consist of constructs to assist in describing the processes of human learning and problem solving in mathematics. Internal representations of mathematical concepts can take various forms, such as individual visualization of mathematics concepts, idiosyncratic notation systems, or attitudes toward mathematics.

External representations, on the other hand, include all external entities or symbols. External representations provide a medium to communicate mathematical ideas, concepts, or constructs. Richard Lesh defined external representations as the embodiment of internal systems of thought. Lesh also referred to external representations as mathematical representations that are simplifications of external systems. Learners use external representations, such as marks on paper, sounds, or graphics on a computer screen, to organize the creation and elaboration of their own mental structures. Unlike internal representation systems, external representation systems can be easily shared with and seen by others.

Multiple Representations in Mathematics Education

In mathematics education, there has been a shift from classic to nontraditional teaching and learning practices with multiple representations, where educators use various representations to effectively present information. Multiple representations refer to different kinds of representations that present the same mathematical ideas from different perspectives or representations that present different aspects of the same mathematical concept. For example, teaching fractions concepts using multiple representations may involve presenting fractions in real-life contexts such as partitioning a pizza or a pie, allowing students to explore equivalent fractions using kinesthetic or virtual manipulatives, or providing students with pictorial representations of fraction operations in addition to formal mathematical representations. Teaching and learning with various kinds of representations provide students with hands-on and minds-on experiences and support a better understanding of mathematical concepts. Also, using multiple representations in mathematics education can help to alter the focus from a computational or procedural understanding to a more comprehensive understanding of mathematics using logical reasoning, generalization, abstraction, and formal proof. A substantial amount of research has demonstrated the effectiveness of multiple representations in enhancing students' conceptual understanding of mathematical concepts.

The notion of multiple representations in mathematics education commonly refers to external representations. However, one of the essential goals of mathematics education is to develop internal representation systems that interact well with external representation systems. James Kaput identified five interacting types of internal and external representations: (1) mental representations—internal representation—that learners construct by reflecting on their experiences; (2) computer representations that model mental representations through computer programs, which allow for arrangement and manipulation of information; (3) explanatory representations consisting of models or analogies that create the interaction between mental and computer representations; (4) mathematical representations, where one mathematical structure is represented by another mathematical structure; and (5) symbolic representations, such as formal mathematical notations.

To understand James Kaput's taxonomy of internal and external representations, consider the different types of representations related to the concept of "slope." When learning about positive slopes, a student might internally imagine a hill, which constitutes an internal (or mental) representation. This mental representation can be replicated on a computer screen. The student can create a unique model that incorporates the mental representation through a computer representation. If the model is viable, then it can be an explanatory representation for the concept of "slope." The student, then, can sketch a similar mathematical graph of the hill and can name the steepness of the hill with the mathematical notation, "slope." This graphical representation of slope can, then, provide support to represent the slope in a symbolic form as a rate of change ($y = mx + b$, where slope is represented with m and indicates the ratio of change on the y-axis to the change on the x-axis). As portrayed in this example, internal and external representations are not separate. Rather, they are intrinsically connected, and they interact continuously. Furthermore, a concept like slope is itself a type of alternative representation. In calculus, a curve is represented by the changing nature of its tangent vector, where the solution to the first derivative at a particular point is the slope of the tangent vector.

Translational Skills Among Different Modes of Representations

In addition to the importance of the effective interactions between internal and external representations in the acquisition and use of mathematical knowledge, it is essential that students develop fluency among different external representations. Richard Lesh enumerated multiple modes through which representations can be constructed: manipulatives, pictures, real-life context, verbal symbols, and written symbols. To demonstrate deep understanding of mathematics, students need to represent their mathematical ideas with different modes of representations and smoothly translate within and between those modes. For example, in algebra, students should be able to make the connection between graphical and algebraic or symbolic representations of equations. Similarly, students need to link what they learn using concrete or virtual manipulatives to both pictorial representations and abstract symbols. For

Children learn groups of five with their hands or may develop proportional thinking as they relate five fingers to one hand and 10 fingers to two hands. (Photos.com)

instance, students who initially learn fraction operations using concrete or virtual manipulatives should be able to relate this knowledge when they later on learn fraction operations using symbolic and more abstract mathematical representations. Connecting different modes of representation simultaneously has been demonstrated to improve conceptual understanding as well as positive attitudes toward mathematics.

In mathematics education research, there is strong evidence that students can grasp the meaning of mathematical concepts by experiencing different mathematical representations and making connections and translations between these modes of representations. Using translational skills among different representational modes encourages students not to merely memorize theorems and facts but also to think analytically to reproduce and use them in real life problems or even in pure mathematical problems.

To deepen students' understandings, teachers should provide students with multiple representations of a single mathematical concept and focus on students' transition ability from one representation to another. Teachers need to be able to present one concept in multiple modes without relying on a single mode and provide students with appropriate transitions among these representations. Teachers should provide also students with ample opportunities to represent mathematical concepts in multiple ways and to connect these representations, thereby developing representational fluency. For example, asking a student to restate a problem in unique words, to draw diagrams to illustrate the concept, or to act out the problem are some ways to provide students with opportunities to translate among representations. If teachers fail to implement the transitioning among different representations, students will be less likely to see how different representations are related and will be more likely to develop misconceptions.

Multiple modes of representation can be used by teachers and students to enhance understanding of mathematics. Most research has shown that providing students with accurate representations improves student learning. However, different representational modes might have different impacts on student understanding. One mode might be more relevant or effective than another for teaching a specific concept. Or, some representational modes can be more appropriate at different developmental stages of the same concept.

For example, research on teaching and learning of fractions has shown that students should be given the opportunity to develop mental representations of fractions using manipulatives before they are presented with symbolic representations. Thus, in addition to using multiple representations, choosing effective and appropriate presentations of information is crucial in teaching and learning. Representations that allow students to actively interact with the subject matter are more effective in student learning than representations that do not support students' active involvement.

Despite the research support for development of higher order thinking skills afforded by different representational forms, little is understood about how students interact with multiple representations in various learning environments. Even though each representation provides similar information, the strain that each representation puts on students' cognitive resources may differ. Not only do individual representations have different impacts on students' conceptual understanding but integrating multiple representations may also result in interaction effects among different modes presented. Therefore, integration of multiple representations becomes an important consideration in the design of instructions. Educators should employ caution as they integrate different modes into instruction, because delivering redundant information with different modes might interfere with learning.

Mathematical Thinking and Representations in the Twenty-First Century

An increasing number of daily activities in the twenty-first century require familiarity with mathematical representations and mathematical thinking. Mathematical thinking, which is a crucial tool for every member of society, includes skills such as pattern recognition, generalization, abstraction, problem solving, proof, and analytical thinking. Most companies prefer employees who are equipped with mathematical literacy or general mathematical skills. However, many students either do not necessarily understand these qualifications or do not value them enough. It is important to emphasize that all humans use mathematical thinking tools in their every day lives and workplaces, with or without noticing they are doing so.

It is not very hard to realize the extent to which mathematical representations are integrated into mundane objects and activities. Consider the number of newspaper columns that provide their readers with different kinds of mathematical representations to explain current issues. Topics in such columns include sports, economics, advertisements, and weather reports. For example, the growth of players, the statistics and ranking of teams, and teams' transfer budgets are represented in several representational modes, such as tabular data, textual information, visual representations, or graphical interpretation. Not only do sports fans need to understand the mathematical information provided readily to them but they also may need to use the mathematical information in problem solving situations, such as estimating the chances of their team's victory. More surprisingly, when a rivalry game is present, the provided data get even more complicated to analyze the chances of each team.

Even though the use of mathematical representations and information in economic and weather columns in various modes is apparent, the ones used within advertisements or political columns may be overlooked. Understanding the mathematical information included in advertisements and deciding which product to buy requires effective use of mathematical thinking tools. In most advertisements, companies present several payment options with different price ranges instead of giving just one price for a product. In particular, mortgage plans to buy houses and installment plans to buy cars require serious analyses of options to choose the best for a given budget. In political columns, on the other hand, one would not be surprised to see percentages representing the proportion of the population that supports various political parties in a country or the votes of a poll. Such information is not only presented as tabular data, visual charts, or graphs, but also as textual information, which is another mode of mathematical representation.

Representations in Problem Solving

Problem solving is one of the essential tools for mathematical thinking. A person equipped with problem solving skills does not necessarily need to have the knowledge base for the solution to each problem encountered but needs to know how to approach problems, locate and access information from different resources, and process information to solve the problem. For example, when one faces a novel problem, an approach to solving that problem can be forming an analogy between the new problem and another, previously solved problem. In other words, known information from an earlier problem can be mapped onto the novel problem. Brainstorming may be another valuable approach to gather different ideas on solution paths to unfamiliar problems. If a problem is too complex, problem solvers can try to break it down into more manageable parts (more solvable problems). One approach to problem solving is solving the problem step-by-step and taking an action at each step to get closer to the goal. Another solving approach can be conducting extensive research to analyze existing ideas and then adjusting possible solutions to the problem in hand. Finally, trial-and-error may be an approach to find a solution to an existing problem. It is emphasized in problem solving that there are many solution paths to a problem and a willingness to try multiple approaches is

---

Representational Skills

The National Council of Teachers of Mathematics presents representation as an important skill needed for students and teachers in teaching and learning mathematics in Principles and Standards for School Mathematics. Students should lucidly and coherently be able to express mathematical ideas through various representational modes, especially in writing and speaking. Through representational skills, abstract concepts can be manipulated into concrete concepts. Developing appropriate representation manipulation skills is necessary to improve conceptual understanding. Further, using various modes of representations, such as graphics, tabular data, mental images, physical objects, mathematical symbols and notations, drawings, and textual information, provides students with organizational skills to systematize their thinking and approach a concept from multiple views, leading to a more coherent understanding. With this ability, students can represent phenomena in a way that is meaningful to them. More importantly, the capability of representing a concept in numerous modes eliminates possible communication problems.

encouraged. Multiple approaches and strategies may be available and some of these approaches may be more efficient than the others.

Problem solving in mathematics, and in other fields as well, requires both knowledge of different representational systems and representational fluency that enables flexible use of various representational systems. For example, when solving a mathematical problem that asks how many quarters there are in 2 1/2, various strategies that involve different representations exist to approach the problem. A student may choose to translate this problem, which is represented in words, into a real-life context, such as how many quarter slices of pizza there are in 2 1/2 pizzas. Another student may opt to draw a picture that represents the given problem and solve the problem using the pictorial representation. Or, some students may represent the problem using symbolic representations and solve the problem accordingly. There may be other approaches where students start with a real-life context and then translate it to a pictorial representation, or where students come up with various relevant representations and choose the most efficient one for them. In more complex problems, different parts of the problems may require different representations. Thus, representational fluency is an essential part of problem solving.

Problem solving is such an important skill that is not only required to help students solve mathematical problems but also provides them with necessary tools to approach and solve problems in the real world. Because the real word does not have recipes to solve a problem, and problem solving requires structured, thoughtful, and careful analysis of problems (especially ill-defined problems) in various situations, people equipped with problem-solving skills are highly valued by employers.

Mathematics as a Language

Mathematics is, to some extent, a language that is universal and can be understood in any part of the world without much difficulty. The mathematics language, which consists of both symbolic and verbal languages, has evolved as the most efficient medium to communicate mathematical ideas and information. Mathematics language also includes graphical images to effectively communicate mathematical concepts and ideas. Thus, different representational modes are used in communicating mathematical ideas and concepts. For example, when a mathematics teacher writes an equation and explains the equation in spoken language to a class, both verbal and written representational forms are in play. Communication in mathematics often involves a constant representational translation between symbolic and verbal representations. Symbolic and verbal languages of mathematics help to express ideas in a meaningful and efficient way. The evolution of mathematics language has been in progress for thousands of years. The goal of this progress is to improve the efficiency of communication, which is central to learning and using mathematics.

Before the emergence of mathematical notations and symbols, mathematicians found it difficult to share their knowledge with the community, even with other mathematicians. Even if a mathematician were able to prove a theorem, for example, geometrically without using mathematical notations and symbols, the mathematician might not have easily written down the proof to share it with others. Difficulties in representing mathematical ideas (writing in a concise and meaningful way using various mathematical notations and symbols) forced mathematicians to seek alternative (especially short and easy) forms to present their knowledge. The need for an effective and efficient mode of communication to convey mathematics ideas resulted in the development of the symbolic mathematical language.

Although the symbolic mathematical language is universal, the verbal mathematical language differs across societies or cultures. For example, although the American and the Japanese use the same symbolic notations to convey mathematical ideas, the verbal language each of these nations uses to communicate about mathematics is different. Differences in verbal languages to communicate mathematics have implications for teaching and learning mathematics. Verbal languages that are clearer about mathematical terms or that relate better to mathematical entities or ideas can support mathematical understanding. For example, counting in the verbal Chinese language is based on the concept of base-10 system. In Chinese, the number 11 is not an arbitrary word in the verbal language. Rather, in Chinese, 11 is "ten-one," 12 is "ten-two," 21 is "two-ten-one," 22 is "two-ten-two," and so on. In other words, the Chinese verbal language clearly conveys that there is one 10 and one 1 in 11 or there are two 10s and one 1 in 21. Such a clear relation between mathematical ideas and verbal language can be an important cognitive tool that supports mathematical understanding.

Further Reading

Curtis, Charles. *Pioneers of Representation Theory: Frobenius, Burnside, Schur, and Brauer*. Providence, RI: American Mathematical Society, 2003.

Goldin, Gerald A. "Representation in School Mathematics: A Unifying Research Perspective." In *A Research Companion to Principles and Standards for School Mathematics*. Edited by J. Kilpatrick, W. G. Martin, and D. Schifter. Reston, VA: National Council of Teachers of Mathematics, 2003.

Kaput, James. "Representation Systems and Mathematics." In *Problems of Representation in the Teaching and Learning Mathematics*. Edited by C. Janvier. Oxfordshire, England: Erlbaum, 1987.

Lesh, Richard. "The Development of Representational Abilities in Middle School Mathematics." In *Development of Mental Representation: Theories and Applications*. Edited by I. E. Sigel. Oxfordshire, England: Erlbaum, 1999.

Lesh, Richard, Kathleen Cramer, Helen M. Doerr, Thomas Post, and Judith S. Zawojewski. "Using a Translational Model for Curriculum Development and Classroom Instruction." In *Beyond Constructivism: Models and Modeling Perspectives on Mathematics Problem Solving, Learning, and Teaching*. Edited by R. Lesh and H. M. Doerr. Oxfordshire, England: Erlbaum, 2003.

National Council of Teachers of Mathematics. "Illuminations: Resources for Teaching Math." http://illuminations.nctm.org.

Utah State University. "National Library of Virtual Manipulatives." http://nlvm.usu.edu/en/nav/siteinfo.html.

Serkan Ozel
Zeynep Ebrar Yetkiner Ozel

Scales

Category: Arts, Music, and Entertainment.
Fields of Study: Algebra; Measurement; Number and Operations; Representations.
Summary: Musical scales have distinct mathematical properties and patterns.

Western music is based on a system of 12 pitches within each octave. The interval between adjacent pitches in this 12-tone system is called a "half step" or "semitone." Pitches separated by two successive semitones are said to be at the interval of a "whole step," or a "tone." Based on a variety of theoretical underpinnings, the concept and sound of tones and semitones have evolved throughout the history of Western music. In modern music practice, a uniform division of the octave into 12 equally spaced pitches, known as "equal temperament," holds sway. Scales are arrangements of half and whole step intervals in the octave. Denoting a half step as h and a whole step as w, the familiar diatonic major scale is defined by the sequence $wwhwwwh$. The diatonic natural minor scale is $whwwhww$. Beginning these patterns from each of the 12 pitches results in 24 distinct diatonic scales. This suggests a set-theoretic description by which each major scale can be represented as a transposition (in algebra this would be called a "translation") of the set of pitches C, D, E, F, G, A, B, and C. In the twentieth century, such mathematical formalisms have led to the conceptualization of non-diatonic scales with special transposition properties.

Octave Equivalence

The concept of octave (the musical interval between notes with frequencies that differ by a factor of two) is fundamental to understanding musical scales. In Western music notation, pitches separated by an octave are given the same note name. The piano keyboard provides a visual representation of this phenomenon. Counting up the white keys from middle C as "1," the eighth key in the sequence is again called C. This eight-note distance explains the etymology of the word "octave." The perception and conceptualization of such pairs of pitches as higher or lower versions of the same essential pitch is called "octave equivalence." Octave equivalence is thought to be common to all systematic musical cultures. Evidence of octave equivalence is found in ancient Greek and Chinese music. Recent psycho-acoustic research suggests a neurological basis for octave equivalence in auditory perception.

The mathematical explanation of octave equivalence comes from the fact that the sound of a musical pitch is a combination of periodic waveforms that can be modeled as sinusoidal functions of time. In the two periodic functions, $f(t) = \sin(t)$ and $g(t) = \sin(2t)$, with frequencies 2π and π, every peak of the lower frequency function coincides with a peak of the high-frequency function. In sonic terms, this is the highest degree of consonance possible for two pitches of different frequencies.

History of Scales

As Western music developed from the Middle Ages through the twentieth century, the central construct was the diatonic scale. This arrangement spans an octave with seven distinct pitches arranged in a combination of five whole steps and two half steps. Interestingly, the pattern of intervals (and not the absolute pitch of the starting note) was the only distinguishing feature of scales until the rise of tonal harmony in the seventeenth century. Pitch-specific examples help illustrate the interval patterns.

The diatonic scale traces its origins to the ancient Greek *genus* of the same name, referring to a particular tuning of the four-stringed lyre (tetrachord) consisting of two whole steps and one half step in descending succession. An example of this tuning can be constructed with the pitches A, G, F, and E. Concatenization of two diatonic tetrachords [A-G-F-{E}-D-C-B] produces the pitches of the diatonic scale (the piano white keys). In medieval European musical practice, the distinct Church Modes (such as Lydian or Phrygian) developed from the diatonic scale by the assignment of a tonal anchor or final tone. For example, the Dorian mode is characterized by the sequence of ascending half and whole steps in the diatonic scale *whwwwhw*; for example D-E-F-G-A-B-C-D, while the Phrygian mode is *hwwwhww*: E-F-G-A-B-C-D-E. The diatonic major scale *wwhwwwh* (C-D-E-F-G-A-B-C) came into widespread use in the seventeenth century. The diatonic natural minor scale is *whwwhww* (A-B-C-D-E-F-G-A).

Intervals, Ratios, and Equal Temperament

The simplest musical interval is the octave. The frequency ratio between pitches separated by an octave is 2:1. The interval of a perfect fifth has frequency ratio 3:2. Using these two ratios, pitches and corresponding intervals for the diatonic scale can be assigned according to Pythagorean tuning. Simpler diatonic scales based on ratios of small integers are known as "just tunings." Western music in the modern era uses a symmetric assignment of intervals known as "equal temperament." In equal temperament, the 12 half steps that comprise the frequency doubling octave each have frequency ratio $2^{1/12} \approx 1.0595$. For these three tuning schemes, frequency ratios relative to the starting pitch and intervals between adjacent scale notes are illustrated and compared in Table 1. For each intonation, the first row gives the frequency ratio from the tonic C to the given note. The second row in each case gives the frequency ratio between adjacent diatonic pitches.

Modern Scales

In contrast to the idiosyncratic pattern of intervals that comprise the diatonic scales, the chromatic scale *hhhhhhhhhhhh* is perfectly symmetric. In particular, the set of pitches that form the chromatic scale is unchanged by transposition—there is only one set of pitches with this intervallic pattern. This set of pitches is referred to as having order one. The elements of the pitch set forming a diatonic scale, which generates 12 diatonic scales by transposition, has order 12. This point of view suggests other scales of interest with respect to transposition. The set of six pitches in a whole-tone scale *wwwwww* (for example, C-D-E-F♯-G♯-A♯-C) are unchanged by transposition by an even number of half steps. A transposition by an odd number of half steps results in the whole tone scale containing the remaining six pitches (for example, C♯-D♯-F-G-A-B-C♯). Thus, the set of pitches in the whole-tone scale has order two. Whole-tone scales are a characteristic feature in much of the music of Claude Debussy.

The twentieth-century composer and music theorist Olivier Messiaen codified a number of eight-tone "scales of limited transposition." Among these are the order three scales *hwhwhwhw* and *whwhwhwh*, which are called "octatonic scales" in the music of Stravinsky and sometimes referred to as "diminished scales" in jazz performance. It can be seen that transposition by one and two half steps produce new diminished scales, but transposition by three half steps leaves the original set of pitches unchanged.

Further Reading

Grout, Donald Jay. *A History of Western Music*. New York: Norton, 1980.

Hanson, Howard. *Harmonic Materials of Modern Music: Resources of the Tempered Scale*. New York: Appleton-Century-Crofts, 1960.

Johnson, Timothy. *Foundations of Diatonic Theory: A Mathematically Based Approach to Music Fundamentals*. Lanham, MD: Scarecrow Press, 2008.

Pope, Anthony. "Messiaen's Musical Language: An Introduction." In *The Messiaen Companion*. Edited by Peter Hill. Portland, ME: Amadeus Press, 1995.

Sundberg, Johan. *The Science of Musical Sounds*. San Diego, CA: Academic Press, 1991.

Eric Barth

Sculpture

Category: Arts, Music, and Entertainment.
Fields of Study: Geometry; Representations.
Summary: Mathematics may be necessary to assure the stability of a sculpture and sculptures can represent mathematical concepts in three dimensions.

The word "sculpture" comes from Latin *sculpere*, meaning "to carve." Sculptures can be made from variety of materials, including wood, metal, glass, clay, textiles, or plastic that is carved, cast, welded, cut, or otherwise formed into shapes. Topiary and bonsai are living sculptures. Modern sculptors even experiment with light and sound. Additionally, sculptures may be free-standing objects or appear as reliefs on surfaces like walls.

The Taj Mahal, one of the most recognizable structures on Earth, includes many geometric reliefs. Sculptures can be static or kinetic, like Rube Goldberg contraptions, and projection sculptures change appearance when viewed from different sides. The outdoor Penrose tribar sculpture in East Perth, Australia, appears to be the illusory figure developed by Roger Penrose when viewed from the correct angle. While mathematical forms have long been used to create sculpture, mathematicians have come to embrace this incredibly flexible art form to investigate many mathematical concepts that might otherwise be difficult to visualize. Many mathematical sculptures are quite aesthetically pleasing in addition to being highly functional in clarifying and representing mathematical ideas. Displays of mathematical sculptures are now a regular part of many art exhibitions and mathematics conferences.

Mathematical Sculptures

Researchers who explore higher degrees of dimensionality often find it challenging to represent these concepts to people whose everyday perception is three-dimensional. Mathematician Adrian Ocneanu's work includes modeling regular solids mathematically and physically. His "Octatube" sculpture, on display in Pennsylvania State University's mathematics building, maps a four-dimensional space into three dimensions using triangular pieces bent into spherical shapes. "Octatube" is conformal; the angles between faces and the way the faces meet are uniform. It was sponsored by Jill Grashof Anderson, whose husband was killed on September 11, 2001. Both graduated with mathematics degrees in 1965. Mathematician Nigel Higson said, "For professionals the sculpture is very rich in meaning, but it also has an aesthetic appeal that anyone can appreciate. In addition, it helps to start conversations about abstract mathematical concepts—something that is generally hard to do with anyone other than another expert."

Other concepts explored by mathematical sculptures include minimum variation surfaces, such as spheres, toruses, and cones, which humans tend to judge to be aesthetically pleasing because of their constant curvature; zonohedra, a class of convex polyhedra with faces that are point-symmetrical polygons, such as parallelograms; and Möbius loops, Klein bottles, and Boy's surfaces, named for mathematicians August Möbius, Felix Klein, and Werner Boy. Sculptures on exhibit at the Fermi National Laboratory, like "Monkey-Saddle Hexagon," focus in part on saddle-shaped minimal surfaces.

Mathematicians Who Sculpt

Art and mathematics have been intertwined for centuries and many historical sculptors such as Leonardo DaVinci were also mathematicians. Cubist sculptors explored many new perspectives on dimension and geometry. Spouses Helaman and Claire Ferguson have created and written extensively about mathematical sculpture. Helaman developed the PLSQ algorithm for finding integer relationships, considered by many to be among the most important algorithms of the twentieth century. He creates his award-winning sculptures to represent mathematical discoveries, and the pair's worldwide presentations have been praised for their accessibility and for initiating dialogue among multiple disciplines.

George Hart, another mathematician-sculptor, has worked in fields like dimensional analysis. He regularly hosts "sculptural barn raisings," where people are invited to help assemble large mathematical sculptures and discuss their properties. This includes a traveling sculpture for use at schools and conferences. Hart also uses rapid prototyping technology for mathematics and sculpture work. In 2010, he left Stony Brook University to be chief of content at the interactive Museum of Mathematics, with an opening date of 2012.

Computer-Generated Sculpture

Self-taught artist and mathematician Brent Collins and computer scientist Carlo Séquin created their Fermi mathematical sculpture exhibit as part of a prolific ongoing collaboration. Séquin started researching geometric modeling in the early 1980s and Collins created saddle-form sculptures during the same period, though he only later learned their mathematical names. The Séquin-Collins Sculpture Generator combines the aesthetics of sculpture, mathematical theory, and computer visualization to allow sculptors to rapidly prototype and refine ideas electronically before beginning to work in their chosen medium. A designer can move around and through the model as well as slice and transform it. Some consider the computer images themselves to be "virtual sculpture." In contrast, some sculptors see computer modeling as too restrictive on the symbiotic processes of design and implementation. Some directions of mathematical sculpture include knots, three-dimensional tessellations, surfaces defined by parametric equations, fractal structures, and models of complex natural entities such as organic molecules.

Other Representations and Projects

The Hyperbolic Crochet Coral Reef project combined mathematics and marine biology to call attention to global warming and other environmental issues using three-dimensional crocheted sculptures of reef lifeforms. Artists create reef components using iterative patterns, which can be permuted to produce a broad variety of lifelike designs. The project is an extension of the hyperbolic crochet work pioneered by mathematician Daina Taimina, who demonstrated that hyperbolic surfaces can be modeled physically.

Some mathematically themed sculptures represent the connections between mathematics and other aspects of society rather than trying to model explicit mathematical concepts. Oakland University's Department of Mathematics and Statistics has a sculpted ceramic mural called *Equation*, which was created to explain the development of mathematics and its relationship to the universe and humanity. Though not a mathematician, artist Richard Ulrish stated that he has fond memories of the mathematics courses he took at Oakland.

Further Reading

Abouaf, Jeffrey. "Variations on Perfection: The Sequin-Collins Sculpture Generator." *IEEE Computer Graphics and Applications* 18 (November/December 1998).

Ferguson, Claire, and Helaman Ferguson. *Helaman Ferguson: Mathematics in Stone and Bronze*. Erie, PA: Meridian Creative Group, 1994.

George W. Hart. "Geometric Sculpture." http://www.georgehart.com/sculpture/sculpture.html.

Hyperbolic Crochet Coral Reef. http://crochetcoralreef.org.

Peterson, Ivars. *Fragments of Infinity: A Kaleidoscope of Math and Art*. Hoboken, NJ: Wiley, 2001.

Maria Droujkova

Six Degrees of Kevin Bacon

Category: Friendship, Romance, and Religion.
Fields of Study: Algebra; Geometry; Number and Operations.
Summary: Concepts from graph theory help explain the idea that people, including actor Kevin Bacon, are surprisingly closely connected with each other.

Six degrees of Kevin Bacon is an example of a network showing a high level of interconnection, known as the "small world" phenomenon. In the language of graph theory applied to films, nodes are film actors, and two nodes are connected by an edge if the corresponding actors have appeared together in a film. It is also a game that tests cinematic knowledge. The task is to find the shortest connection between a given actor and Kevin Bacon. For example, John Wayne is two connections from Kevin Bacon. They were never in a film together, so the distance is greater than one. John Wayne starred with Eli Wallach in *How the West Was Won*, and Eli Wallach starred with Kevin Bacon in *Mystic River*, establishing a shortest distance of at most length two.

The idea of quantifying distance by interpersonal connections dates at least to a 1929 short story called *Chain-Links* by the Hungarian writer Frigyes Karinthy, wherein the narrator determines a five-step connection between a riveter at the Ford Motor Company and himself. Almost 40 years later, the social psychologist Stanley Milgram, best known for his experiments on obedience to authority, devised an experiment to quantify interpersonal connections empirically. Letters were given to some 300 participants, each charged with forwarding the letter to an acquaintance who should move the letter toward the intended recipient. Writing in 1969 with Jeffrey Travers, Milgram stated, "The mean number of intermediaries observed in this study was somewhat greater than five; additional research (by Korte and Milgram) indicates that this value is quite stable." Rounding up, this value became the popular notion "six degrees of separation"—that any two people on the planet are connected by six links. It served as the title of John Guare's 1990 play and 1993 movie about the confidence man David Hampton. In the play, a character speaks to the audience, "Six degrees of separation. Between us and everybody else on this planet. The President of the United States. A gondolier in Venice. Fill in the names. I find that A) tremendously comforting that we're so close and B) like Chinese water torture that we're so close." Exactly how close people are is something sociologists continue to debate, since the nodes and edges of this network are not precisely known.

Mathematics Networks

There are large networks where the nodes and connections are exactly known, allowing for precise analysis. In a collaboration network, nodes are researchers, and two nodes are connected by an edge if the corresponding researchers worked together on a published paper. As early as 1957, mathematicians determined their "Erdös numbers," the collaboration distance from Paul Erdös, the most prolific mathematician of recent years, with some 1500 published research papers and more than 500 collaborators. For instance, the author never wrote a paper with Erdös, but Robin Wilson wrote a paper with Erdös in 1977, and the author wrote a paper with Robin Wilson in 2004, so the author's Erdös number is two. The American Mathematical Society's MathSciNet electronic publication computes the "collaboration distance" between any two authors in its database of some 500,000 authors and 2.5 million publications.

Film Networks

Of more interest to the general public than mathematicians and their papers, the Internet Movie Database (IMDb, found at imdb.com) includes over 1 million actors around the world and some 250,000 films from the 1890s to titles in production. The Web site Oracle OfBacon.org accesses the IMDb and determines the shortest link between any two actors. The network is very tightly connected; it is surprisingly difficult to name any pair of actors even four apart. Consider Kevin Bacon, who has been in over 60 films with over 2200 total co-stars. That is a very small percentage of the total number of actors in the database, but there are over 225,000 actors who, like John Wayne, are co-stars of co-stars of Kevin Bacon. Actors within four links of Kevin Bacon comprise approximately 98% of the database. About 99% of the actors in the IMDb all connect to one another. Finding actors within the last 1% who are five or more from Kevin Bacon is another entertaining part of the game. As of 2010, there are 17 actors with a distance of eight from Kevin Bacon, so that "six degrees" is a misnomer.

Another variant of the game is to determine the actor who is best connected on average. The average every actor's Kevin Bacon number is 2.980. This number means, roughly, that a randomly chosen actor is within three links of Kevin Bacon. It is interesting to consider which sorts of actors have the lowest averages. John Wayne, with significantly more movies and co-stars than Kevin Bacon, has an average of 3.026 links to the rest of the connected actors. The best-connected actor, as of 2010, is Dennis Hopper, with an average distance of 2.772. The IMDb is regularly updated with new actors and films, and the connection data change accordingly.

Why is it six degrees of Kevin Bacon, and not some other actor? The game was created by students at Albright College in January 1994; they had watched *Footloose* earlier in the day, then saw a commercial for another Kevin Bacon film, *The Air Up There*, and a pop culture phenomenon was born. There are similar games based on other large databases, such as baseball players connected by teams, and "six degrees of" remains a very common phrase in society. Kevin Bacon himself used the notion to build a Web-based charity fundraiser, SixDegrees.org. The notion of "small world" networks is being used by scientists in applications as diverse as neural networks of worms, the interconnection of power grids, analysis of the World Wide Web, and genealogical connections.

Further Reading

Grossman, Eric. "The Erdös Number Project." *Oakland University*. http://www.oakland.edu/enp/.

Hopkins, Brian. "Kevin Bacon and Graph Theory." *PRIMUS* 14, no. 1 (2004).

Watts, Duncan. *Six Degrees: The Science of a Connected Age*. New York: Norton, 2003.

Brian Hopkins

Skating, Figure

Category: Games, Sport, and Recreation.
Fields of Study: Algebra; Geometry.
Summary: The elements, equipment, and scoring system of figure skating all involve a mathematical framework.

Figure skating is a winter Olympic competitive sport, which involves artistically gliding on ice using metal blades. Ice skating rinks are generally shaped in the form of rectangles with rounded corners. The patterns skaters form on the ice can be explained in geometric terms. Physical principles are observed when watching figure skating. The scoring system used to judge figure skating involves algebraic computations.

Patterns

The bottoms of ice skating blades are not flat, but rather slightly curved, like arcs taken from the edge of a circle about seven to nine feet in radius. This enables the skater to angle and tilt to form patterns on the ice. These patterns can be represented geometrically. For instance, the most famous geometric pattern on ice is a figure eight, which can be formed by two circles of equal size tangential to each other. A skater could start the first circle of the figure eight on the right forward outside edge and skate the second circle on the left forward outside edge. The possible edge combinations include using the left or right foot, traveling forward or backwards, and using the inside or outside edges.

Mathematical Principles of Spinning and Jumping

In addition to basic compulsory figures, modern skating requires participants to execute increasingly difficult jumps and spins. In a jump, the skater's center of gravity follows a parabolic arc with respect to the ice, and a jump is frequently measured in terms of its vertical displacement (the height off the ice) as well as horizontal displacement (the distance). Both are a function of many variables, such as the takeoff angle and velocity immediately prior to the jump.

Spinning, whether in the air as part of a jump or on the ice, is also a complex function of many variables. Factors include the skater's body mass and speed when entering the spin, as well as the extension of the arms or legs from the body. For example, a spinning skater rotates more slowly with extended arms than when the arms are tucked in because as the radius between the body and the arms decreases, the angular velocity increases.

Judging

Four disciplines of figure skating are competitive at the Olympic level: singles (ladies' and mens'), pairs, and ice dance. In each of these disciplines, a choreographed program is skated to music in competition and is judged according to the International Skating Union's International Judging System. The International Judging System awards points for technical difficulty and artistry.

There are many types of skating elements. Jumps vary from their takeoff edges as well as numbers of rotations between one and four. Throw jumps are also performed by the pair teams. A variety of spins are possible, but there are three basic spin positions: upright, camel, and sit. Some spins involve a change of foot, change of position, flying entrance, or difficult variation. Footwork is an element in every program and requires steps and turns that fully cover the ice surface in a circular, straight line, or serpentine pattern. For pairs and ice dance skaters, combination spins, lifts, and other elements requiring two skaters are also scored.

Each of the skating elements performed in a program is assigned a numerical base value, which varies according to difficulty. For example, in the 2010–2011 skating season, the base value of a triple toe loop was 4.1 points, and the base value of the single toe loop was 0.4 points, indicating that the triple toe loop was a much harder jump. Judges add to or subtract from the base value of each element depending upon its execution. For instance, a poorly performed toe loop would receive fewer than 0.4 points. The sum of the values given for each element is called the "technical score."

In addition to a technical score for performance on the individual elements, overall scores for artistic aspects of the program, such as choreography, interpretations, transitions, and skating skills, are awarded as the program components score, which is added to the technical score for a total overall score. The skaters with the highest scores earn the highest rankings.

Further Reading

Carroll, Maureen, Elyn Rykken, and Jody Sorensen. "The Canadians Should Have Won!?" *Math Horizons* 10 (February 2003).

Kerrigan, Nancy, and Mary Spencer. *Artistry on Ice: Figure Skating Skills and Style*. Champaign, IL: Human Kinetics, 2002.

Schulman, Carole. *The Complete Book of Figure Skating*. Champaign, IL: Human Kinetics, 2002.

Diana Cheng

Skydiving

Category: Games, Sport, and Recreation.
Fields of Study: Algebra; Calculus.
Summary: Principles of calculus can be used to model a sky dive and to calculate the effect of the parachute on velocity.

Skydiving is the act of leaping out of an airplane at a sufficient altitude and placing your life in the hands of a piece of cloth—although a fairly large piece of cloth. Leonardo da Vinci left drawings of parachutists in his *Codex Atlanticus* circa 1485. The modern parachute was invented by Louis-Sébastien Lenormand in France, making the first public jump in 1783. In 1797, André Garnerin was the first to use a silk parachute, earlier versions being made of linen. The first parachute jump from an airplane was in Venice Beach, California, in 1911. The parachute was held in the arms and thrown out as the jumper left the plane. The soft-pack parachute was developed in 1924. There are two types of parachutes used for skydiving: round, and ram-air (square). The U.S. Army uses the round 35-foot diameter parachute to train its paratroopers because they are reliable and give the jumper a terminal velocity of about 15 feet per second. Most skydivers in the United States started using a 28-foot round canopy. They produced a terminal velocity of about 17–18 feet per second—a somewhat hard landing. The switch to ram-air types came in the 1970s; these give more comfortable landings and maneuverability. The rates of descent vary from canopy to canopy, but terminal velocities usually run from eight feet per second (5.5 mph) to 14 feet per second (9.5 mph).

A canopy's performance is determined by its wing-load, which helps determine the terminal velocity and speed at landing. Most canopies are flown with a wing-load between 0.8 and 2.8 pounds per square foot. To compute the right size of canopy, take the total weight (W) of the jumper and equipment divided by the assigned wing-load factor (WLF):

$$Area_{canopy} = \frac{W_{jumper} + W_{equipment}}{WLF}.$$

To model the parachute jump itself is much more complicated. It involves a first order differential equation to find the speed. The forces on a skydiver are the gravitational force, F_g, and the drag force, F_d, of air resistance and buoyancy. There are two factors to the drag: the time before and the time after the canopy deploys. If x is the distance above the Earth's surface, then $a = dv/dt$ is acceleration and $v = dx/dt$ is velocity. For most jumps, the gravitational force stays essentially constant.

In a first approximation to the problem, take the drag force to be proportional to the velocity. The coefficient of drag has one value when the skydiver is falling and a second value when the parachute is fully deployed. During the fall, the velocity satisfies the initial value problem:

$$m\frac{dv}{dt} = -mg - k_1 v \qquad v(0) = 0.$$

This is a separable ordinary differential equation. Its solution can be found by most students in a calculus class. The jumper's position then is found by integrating the velocity with initial condition that at time $t=0$ the jumper is at the jump altitude. After the chute deploys, the velocity and position can be found exactly as above, except that the drag coefficient and initial conditions change.

A second approach is to assume that the drag force is proportional to the square of the speed. Then, a falling object reaches a terminal velocity:

$$V_T = \sqrt{\frac{2mg}{\rho A C_d}}$$

where V_T is the terminal velocity, m is the mass of the falling object, g is the acceleration due to gravity, C_d is the drag coefficient, ρ is the density of the fluid through which the object is falling, and A is the projected area of the object.

Based on air resistance, the terminal velocity of a skydiver in a belly-to-Earth free-fall position is about 122 miles per hour (179 feet per second). A jumper reaches 50% of terminal velocity after about three seconds and reaches 99% in about 15 seconds. Skydivers reach higher speeds by pulling in limbs and flying head down, reaching speeds close to 200 miles per hour. The parachute reduces the terminal velocity to the five to 10 miles per hour range. This is achieved by increasing the cross-sectional area and the drag coefficient, lowering the terminal speed.

Further Reading

Meade, Doug. "Maple and the Parachute Problem: Modeling With an Impact." *MapleTech* 4, no. 1 (1997).

Meade, Doug, and Allan Struthers. "Differential Equations in the New Millennium: The Parachute Problem." *International Journal of Engineering Education* 15, no. 6 (1999).

Poynter, Dan, and Mike Turoff. *Parachuting*. 10th ed. Santa Barbara, CA: Para Publishing, 2007.

The United States Parachute Association. http://www.uspa.org/.

DAVID ROYSTER

Soccer

Category: Games, Sport, and Recreation.
Fields of Study: Data Analysis and Probability; Geometry; Measurement.
Summary: Mathematical modeling and statistical analysis can help inform individual techniques and team tactics in soccer.

Soccer is a sport that has been enjoyed worldwide for more than a century by both players and spectators. In the early part of the twentieth century, mathematician Harald Bohr, founder of the field of almost periodic functions and brother of famed physicist Niels Bohr, was a skilled player and a silver medalist on the 1908 Danish Olympic soccer team. He was reported to be so popular that his doctoral dissertation defense was attended by more soccer fans than mathematicians. In general, the sport is often cited for its equal emphasis on individual skills and team tactics. As in other sports, statistics are frequently cited by sports commentators. In addition, technically demanding individual actions, as well as masterfully executed plays, can all be described and analyzed using statistics and mathematics, which is done worldwide by numerous sports scientists. One could even say that the players, perhaps unconsciously, use or display "mathematics in motion."

Individual Technique

The effectiveness of any of the various moves a player uses (kicking, heading, or dribbling) depends on a combination of physical qualities and technical skills. This idea can be demonstrated using the instep kick as an example; the instep kick, with the aim to kick the ball as hard as possible, is by far the most studied soccer movement by sport scientists. In order to maximize the forward swinging speed of the shank, physical qualities (such as strength and speed of contraction) of the knee extensor muscles and the hip flexor muscles are important. However, research has shown that technical skills are equally important. The specific technical skill required for optimal kicking is coordination—how the shank moves relative to the thigh.

Coordination is one of the topics studied in the scientific field of biomechanics, which relies heavily on mathematics. Biomechanics researchers use high-speed cameras in their laboratories to record kicking performance from top level players. From the video footage, the researchers can obtain the three-dimensional position in space of selected points on the kicking leg. Using mathematical concepts from vector algebra and trigonometry, joint and segment angles can subsequently be calculated. These data, in turn, allow calculations of a number of kinematic parameters of the foot, shank, and thigh, comprising linear velocity and acceleration and angular velocity and acceleration.

In mathematics, the most common way to calculate velocities and accelerations from position data is to use calculus. This method, however, requires the position data to be specified as a mathematical function. This is not the case with position data obtained from video footage, which are discrete in nature—they consist of thousands of numbers, specifying the three-dimensional position of numerous points on each video frame. From the cameras' frame rate, the elapsed time between frames can be calculated, which instead allows numerical differentiation of the position data using a computer. Finally, by combining the kinematic data with data for each segment's mass and moment of inertia (a measure of a segment's inertia when rotating) and using the principles from Newtonian mechanics, the researchers can calculate how the movement of the thigh affects the movement of the shank and vice versa. The forward swing of the thigh generates a force at the knee that causes the shank to swing faster forwards. The force is larger, the faster the thigh moves, while the effect of the shank is larger, the closer the knee angle is to 90 degrees. Top players instinctively coordinate thigh and shank movements in order to take maximum advantage of these intersegmental forces, although science so far has failed to determine precisely what optimal coordination is.

Team Tactics

When a midfielder executes a beautiful play that a forward picks up between defending opponents and scores, a lot of "hidden" mathematics is occurring. The midfielder's team members and opponents are all moving simultaneously in different directions with different speeds, yet the midfielder still manages to precisely calculate the required ball speed and direction to execute his play, so the ball and forward meet at the intended spot out of reach of defending opponents. Situations like this are analyzed by sport scientists and coaches using the methods of notational analysis. With video footage and specialized software, the various actions

The skill required for optimal kicking is coordination—how the shank moves in relation to the thigh. (Photos.com)

(sprinting, moving sideways, tackling, or heading) of each player from both teams can be registered. Statistical calculations can reveal which situations are most likely to lead to a certain outcome, such as scoring a goal, and which general tactics lead to most of these situations. Digital representations have also been used to assist with tactics and analysis. Researchers from the University of Sheffield digitized a soccer ball (including even the stitching) and computed airflow around the ball. They found that the specific shape and surface of the ball, and its initial orientation, are significant in determining the ball's trajectory through the air. Measurements on actual balls in a wind tunnel at the University of Tsukuba verified these mathematical simulations.

Further Reading

Chartier, Tim. "Math Bends It Like Beckham." *Math Horizons* 14 (February 2007).

Putnam, C. A. "Sequential Motions of Body Segments in Striking and Throwing Skills: Descriptions and Explanations." *Journal of Biomechanics* 26 (1993).

Reilly, T., and M. Williams. *Science and Soccer.* New York: Routledge, 2003.

Henrik Sørensen

Sport Handicapping

Category: Games, Sport, and Recreation.
Fields of Study: Algebra; Data Analysis and Probability; Number and Operations.
Summary: Various calculations are used to set fair, competitive handicaps in sports.

Sport handicapping is an important methodology that affects millions of people worldwide and potentially impacts billions of dollars worth of bets. In many sports, handicaps are calculated for individuals or teams and are used as a way of "equalizing" performance by giving a scoring advantage or other in-game compensation to some players. This process allows lower skilled players to compete with higher skilled players while preserving perceived fairness. The term "handicap" refers to both the adjusted scores and the process of determining them, and may also be used for whole tournaments that rely extensively on the method. Handicap in this context derives from a seventeenth century lottery game called "hand-in-cap," where players put their bets in a literal cap. A point spread, frequently used in sports betting, is a related idea for computing or estimating relative advantage to equalize teams in competitive sports. Examples of sports using handicapping at various levels include bowling, golf, horse racing, and track and field.

Handicapping

In sport, a handicap is usually imposed to enable a more equal competition to take place. The handicap is calculated according to specific criteria set down for each of the sports that use the technique, meaning that some are much more complex than others. To understand why a handicap may be used, consider one of the most well-known sports that employ a handicap system, golf.

If a recreational golf player were to compete against the best golfer in the world in a round of golf, then the outcome would almost certainly be a win for the better golfer. A win by a large margin would also have been very likely. If a handicap were applied that was based on each player's average scores, then the outcome would be much less certain. There would have been a distinct possibility that, if the recreational player had played well, they would have had the opportunity to beat the better golfer—or at least not loose by many shots—after the handicap was applied.

In most sports when professionals compete against each another, the events are usually free from handicapping. A professional golf tournament will usually engage those who play with a scratch (or zero) handicap.

One of the primary reasons for using a handicap is to make an event more competitive. In many respects, this makes the given sport more enjoyable and can help to make it more appealing and increase the number of those wishing to participate.

Tenpin bowling is a sport that has more participants worldwide than most other sports. The overwhelming majority of players are recreational, although many take part in annual league competitions. Most leagues are not scratch based (on actual total pin fall) but are handicapped. In tenpin bowling handicap leagues, the scores that are used to determine who has won are a combination of the total pins actually knocked down and the handicap value. This method allows players (and teams) with lower averages to compete against players (and teams) that have much higher averages.

The handicap in Tenpin Bowling is usually of the form: Handicap value (per game) = 80% of the difference between the player's average and 200 pins.

If a bowler averages 100 pins, then the bowler would, using the handicapping system, gain a handicap value of 80 pins: $(200 - 100) \times 0.80$. The total pinfall for a game would be 80 plus whatever number of pins the bowler actually knocked down.

This handicap system is versatile in that the two values used (the 80% and the 200 pins, in the example above) can be manipulated to suit the particular league. For instance, if there are a number of players who average over 200, for example 210 or 220, then the handicap may be 80% of the difference between each bowler's average and 220 pins. Alternatively, if the players are grouped quite closely together, then the handicap may be 66% of the difference between each bowler's average and 200 pins.

Athletics

Athletics, or track and field, is another mass participation sport, but one in which, at the highest level, age is intrinsically linked to performance—few athletes compete internationally in their late 30s and beyond. There is still huge participation in the sport by people older than 30, and there are obvious health benefits to doing so.

There is a scoring system that takes age into account by comparing race time to that of the world record holder in each age group. It is often known as a World Association of Veteran Athletes (WAVA) Rating and is expressed as a percentage between zero and 100. If one gets a WAVA rating of 50%, it means that the competitor is half the pace of the world record holder. WAVA rating is a useful way to make comparisons between runners of all ages and can form the basis of a handicap league.

Horse Racing

A further important application of handicapping is that seen in horse racing, a sport on which billions of dollars worth of bets are made each year. In a handicapped race, the horse must carry a certain additional weight, which when added to the weight of the jokey gives it an assigned impost (or total weight). These weights are held in saddle pads with pockets.

The calculation for the weight a horse is required to carry is based on a number of factors. A great deal of work is done with past data to create and then ensure that the handicaps are as fair as possible. These handicaps allow for horses of differing abilities to race against each other over a given distance.

Further Reading

Mullen, Michelle. *Bowling Fundamentals*. Champaign, IL: Human Kinetics, 2003.

Tuttle, Joeseph J. *The Ultimate Guide to Handicapping the Horses*. Self published: Createspace, 2008.

Wright, Nick. *Lower Your Golf Handicap: Under 10 in 10 Weeks*. London: Hamlyn, 2006.

Stephen Lee

Square Dancing

Category: Arts, Music, and Entertainment.
Fields of Study: Communication; Geometry; Representations.
Summary: Square and contra dancing employ many mathematical principles, including symmetries and permutations.

Square dance is geometry and combinatorial mathematics in motion. A caller directs the dancers through a set of choreographed dance movements unique to each type of square dancing. The dancers are sorted and shuffled in a myriad of ways by the caller and then returned to their original positions. Not only do the participants create mathematical forms as they move, mathematics is used to analyze different aspects of square dancing and its related form, contra dancing. For example, graph theory, matrix theory, and group theory can be used to represent the various structures and symmetries. Mathematics has also been used to analyze optimal calling patterns depending on the specific combinations of movements in the dance. Square dancing is a popular pastime for many people with an interest in mathematics. Several colleges have square dancing clubs, such as the Square Roots at North Central College in Illinois. That college has also offered a course called "The Mathematics of Square Dancing," which combined advanced dance patterns with discussions of mathematics theory, including parallelogram or hexagon dancing.

The Basic Square

The basic square consists of four couples. A square is symmetric under rotations of 90, 180, 270, and 360 degrees. Some or all of the dancers in the basic square can rotate in a circular movement according to these symmetries. Including the mirror reflections about each of the two lines of symmetry passing through the center of the square and parallel to an edge, there are six different targets of movement for the dancers. Further, in respect to each male (m)-female (f) pair, there are 10 possible movements. Thus, f_1 could be directed to replace either f_2 or m_2, m_1 could replace either f_2 or m_2, or both f_1 and m_1 could replace f_2 and m_2. Since there are four pairs, there are 240 possible movements among the dancers ($6 \times 10 \times 4 = 240$). Dance is about movement and not positions; thus, dance movements are not transitive. A movement of f_1 to f_2 is not the same as a movement of f_2 to f_1, although the outcome is the same arrangement. The two cases differ in respect to who initiates the action and who must react to the other's actions.

Secondary Squares

Besides the basic square, several other squares are part of square dancing. First, each m-f pair is a square. Several calls direct the movements of these dancers relative to one another. Thus, in a Do-Si-Do, the two members dance a square around one another and return to their initial positions. Alternatively, the basic square can be divided into a square within which a circle is inscribed. Four of the dancers constitute the square, while the remaining dancers move inward so that they are contained by the larger square. These can then be instructed to move according to the four symmetries. This arrangement can be inverted. The pairs can move toward a center point and form the radii of a circle, while the square that contains the circle is implicit. Again, the four symmetries constrain these movements. Instead of being expanded, the square can be constricted. The larger square can be divided into two smaller squares, each with four dancers. The dancers can be instructed to form smaller squares with the pair on the right, the pair opposite, or the pair on the left.

Decomposing Squares Into Columns and Lines

Besides arranging dancers in squares and circles, the caller can also arrange them into columns and lines. A column arrangement occurs when all the couples are aligned one behind the other. A caller can shuffle this arrangement into any of 24 possibilities. A column of dancers can then be bisected longitudinally into two lines or crosswise into two smaller squares. There are two kinds of lines: one in which all dancers face the same direction, and another (a wave) in which they alternate the direction they are facing.

Further Reading

Mathematical Association of America. "Square Dancing Takes a Mathematical Spin." http://mathdl.maa.org/mathDL/pa=mathNews&sa=view&newsId=230.

Mui, Wing. "Connections Between Contra Dancing and Mathematics." *Journal of Mathematics and the Arts* 4, no. 1 (2010).

MICHAEL K. GREEN

Step and Tap Dancing

Category: Arts, Music, and Entertainment.
Fields of Study: Communication; Geometry; Representations.
Summary: Step and tap dancing each involve rhythms and combinations that can be analyzed mathematically.

Step dance is the type of dance focusing on feet movements. It de-emphasizes the other two spatial dance aspects—hand and body movement—and repositions dancers relative to the ground to form movement patterns. There are forms of step dancing in several cultural traditions, such as Malambo from Argentina, Irish stepdance, African-American stepping, and traditional Cherokee dancing. Related forms include clog and tap dancing.

The movements of these styles of percussive dance may be performed by a single dancer or choreographed among several dancers. Tony Award–winning choreographer and dancer Danny Daniels noted that, while an individual dancer may improvise, groups must be coordinated. The rhythms and counts for the dances he designed or performed on Broadway could be organized and detailed using mathematically based musical notation. Dance theorist Rudolf Laban used ideas from various fields, including crystallography, when he modeled dance dynamics. Scientists and dancers continue to develop notation and models to express human movement in tap and other dances. Dance algorithms may help create natural robotic movement. Dancer Gregory Hines said: "My style is part choreography, part improvisation. That gives me a chance to show people the possibilities of tap dancing, which, at its heart, is mathematics with endless possibilities."

Ratio and Proportion

There are several ratios related to music and choreography that determine movement in step dancing. Music time signature is written as a fraction with the denominator signifying the size of the notes used, and the numerator signifying the total length—in such notes—of a bar, which is the unit of music. For example, traditional music for Irish slip-jig has 9/8 time signature in the note pattern: quarter, eighth, quarter, eighth, dotted quarter (three-eighth). The five notes in the time signature correspond to two-and-a-half dance steps per bar, with long graceful slides between the steps.

The formula for a dance includes the number of bars in each repeating cycle (sometimes performed symmetrically) first for one starting foot and then the other. For example, a song that has 40 bars may be choreographed to include five step cycles, each spanning eight bars. Another ratio important for step dancing is the tempo of music, measured in beats per minute (bpm). Dancing competitions specify the tempo range for each type of dance. For example, single jig must be 112–120 beats per minute. Tap dancers of the past used their signature "time steps" (particular combinations of taps) to communicate the tempo to the accompanying band.

Patterns and Improvisation

In step dances, themes are expressed using sequences of the basic elements or steps. For example, common elements in tap dancing include shuffles, flaps, pullbacks, wings, and stomps. These sequences may be strictly choreographed from beginning to end, sometimes with repeating patterns or permutations of shorter elements, which can be repeated by any dancer who has learned the sequence. Improvisation allows the dancer to take basic elements and rearrange them in ways that may appear to be random to the casual observer.

Some step dance music has built-in departures from the standard bar structures. For example, Irish stepdance "crooked tunes" may have seven-and-one-half bar parts in addition to eight bar parts. Step dance patterns have multiple levels: steps within a bar, combinations of steps spanning multiple bars, and patterns of these step combinations. Order and perceived randomness can be manifested at all levels.

Dance-Dance Revolution

Dance-Dance Revolution (DDR) is a step dancing video game. The goal of the game is to match the pattern of steps on the screen and their rhythm on the special gaming pad with four or eight foot positions. The combination of visual, audio, and kinesthetic representations of the same rhythm have kept versions of the game popular around the world since its release in 1998.

Later versions of DDR use a mathematical visualization of multi-dimensional data, called radar diagrams, to rate the difficulty of individual dances. The variables describe different characteristics of the dance, such as steam (the density of steps) and chaos (the amount of steps that do not occur on beat).

Further Reading

Apostolos, M. K., M. Littman, S. Lane, D. Handelman, and J. Gelfand. "Robot Choreography: An Artistic-Scientific Connection." *Computers & Mathematics with Applications* 32, no. 1 (1996).

Maletic, Vera. *Dance Dynamics: Effort and Phrasing*. Columbus, OH: Grade A Notes, 2005.

Sethares, William. *Rhythm and Transforms*. New York: Springer, 2007.

Maria Droujkova

String Instruments

Category: Arts, Music, and Entertainment.
Field of Study: Geometry; Number and Operations; Representations.
Summary: The harmonics and timbre of wind instruments are described and computed using mathematics.

All stringed instruments exhibit a fundamental property of physics in that when impacted, they vibrate at numerous frequencies. The vibration of the string displaces the air around it, which—when impacted on the human eardrums—creates the sensation of sound. Some of the common instruments in the string family are violin, guitar, harp, mandolin, cello, and banjo. A modern violin has about 70 parts, and the overall design of such complex string instruments is inherently mathematical. Features such as string tension, area, and shape of the top plate, and spacing of frets all have mathematical properties that influence sound. For any string, at a given tension, only one note will be produced. To generate multiple notes from the instrument, many strings may be used to span the desired frequency spectrum (for example, harps) or the string may be forced to vibrate at different lengths, thereby changing the frequency (for example, guitars). On an equally tempered instrument like a guitar, the spacings of the frets, which help a player adjust string length, have to be scaled by the ratio $2^{1/12}$. This problem is mathematically equivalent to duplicating a cube, which is one of the classic problems of antiquity. Mathematician Jim Woodhouse has studied violin acoustics using linear systems theory and mathematically modeled "virtual violins," as well as related vibration problems like vehicle brake squeal.

Harmonic Series and Fundamental Frequency

When a string is plucked, struck, or bowed, it resonates at numerous frequencies simultaneously. The waves travel up and down the string. These waves reinforce and annul each other, which results in standing waves. The one-dimensional wave equation is used to model string instruments. A harmonic series is composed of frequencies that are an integer multiple of the lowest frequency. Fundamental frequency is the lowest frequency in a harmonic series. The musical pitch of a note is usually perceived as the fundamental frequency. The fundamental frequency (f) of a string can be computed as

$$f = \frac{\sqrt{\frac{T}{\frac{m}{L}}}}{2L}$$

where T is the string tension in newtons, m is the string mass in kilograms, and L is the string length in meters. The fundamental frequency is also known as the "first harmonic."

Timbre

Timbre is the quality of a musical note and is what defines the character of a musical instrument. When two different instruments play the same note, the note could have the same frequency. The human ear distinguishes the source of the note because of timbre. Hermann Helmholtz was the first to describe timbre as a property of sound. When an instrument plays a certain note, the outputted sound consists of the fundamental frequency and its harmonics. These harmonics differ from instrument to instrument—what is known as "timbre."

Further Reading

Hall, Rachel W., and Kresimir Josic. "The Mathematics of Musical Instruments." *American Mathematical Monthly* 108, no. 4 (2001).

Mottola, R. M. "Liutaio Mottola Lutherie Information Website: Technical Design Information." http://liutaiomottola.com/formulae.htm.

Rossing, Thomas. *The Science of String Instruments.* New York: Springer, 2010.

Ashwin Mudigonda

Sudoku

Category: Games, Sport, and Recreation.
Fields of Study: Algebra; Number and Operations.
Summary: The game of Sudoku is explained by and informs graph theory and randomness.

Sudoku is a number puzzle based upon the mathematical concept of a "Latin square." Latin squares are arrays of numbers in which each number is listed only once in any row or column. Leonhard Euler originated the term, calling them "Latin squares" because he used Latin letters rather than numbers in his investigations. Completed Sudoku grids are Latin squares that are further subdivided into subgrids in which the numbers also appear only once in the subgrid. The graphic below shows a completed 9-by-9 Sudoku puzzle.

7	3	1	4	9	5	8	6	2
9	8	4	6	2	1	3	5	7
5	2	6	3	7	8	9	4	1
4	1	9	5	3	2	7	8	6
8	7	5	1	6	9	4	2	3
3	6	2	7	8	4	1	9	5
1	9	8	2	5	7	6	3	4
6	5	7	8	4	3	2	1	9
2	4	3	9	1	6	5	7	8

Originally published in the 1970s in Europe and the United States, Sudoku surged to popularity in Japan in the late 1980s and reappeared in Europe and America in the mid-1990s, becoming popular among puzzlers. The popularity of the puzzle has continued to grow in the twenty-first century, leading the puzzle to become the subject of mathematical scrutiny.

Sudoku is usually based on a Latin square with nine rows and columns; puzzles of other sizes are possible such as 4-by-4, 16-by-16, and 25-by-25. Some of the numbers are filled in to start. The goal is to quickly and accurately complete the puzzle, with the digits 1–9 placed once in each row, column, and subgrid. Below is an unsolved Sudoku puzzle:

	1	3	5		9	2		
2				8				
9	8		6		7		1	
3	5						8	
		6	8		1	7		
	2						9	1
	6		3		4		2	5
				5				9
		2	9		6	8	4	

Graph Theory

There are a number of interesting mathematical questions associated with Sudoku puzzles, in particular the conditions under which they have one solution. In 2007, Agnes Herzberg and M. Ram Murty showed that Sudoku puzzles can be recast as graph coloring problems allowing the broad, well-developed theory of graphs to be applied to the solution question. In particular, they showed that a standard Sudoku puzzle can be thought of as a graph where each cell in the puzzle is represented by a vertex with 20 edges, each edge connecting the cell to another cell in the row, column, or sub-grid. Graphs for which all vertices have the same number of edges are called "regular," so Sudoku graphs made in this way are "20 regular" graphs.

Since each digit can appear only once in any row, column, or subgrid, putting the nine digits into the cells is equivalent to coloring the vertices of the graph with nine colors such that no vertices connected by an edge are the same color—or in graph theory terminology, finding a "proper 9-coloring" of the graph. The number of ways to color a regular graph with n colors is a well-known formula that is a function of the number of colors and the number of vertices. Using this and other ideas about coloring graphs, Herzberg and Murty proved that 9-by-9 puzzles must have at least eight different digits shown in the starting configuration to have a unique solution.

There are still many unanswered questions about when Sudoku puzzles have one solution. Assuming that eight different digits are used in the starting configuration, how many numbers total must be shown in the starting configuration to ensure a unique solution? The answer is not known; a small number of distinct puzzles with 17 entries in the starting configuration are known to have a unique solution. There are no known puzzles with 16 or fewer entries that have unique solutions. Does one exist? Would the answer be different if all nine digits are used in the starting configuration? Mathematicians and puzzlers are investigating these and other interesting questions.

For example, in 2010, mathematicians Paul Newton and Stephen DeSalvo demonstrated that the arrangement of numbers in Sudoku puzzles is more random (by some definitions of randomness) than 9-by-9 matrices produced by random generators, since Sudoku rules excludes some of the possible arrangements that have innate symmetry.

Further Reading

Chevron Corporation. "Sudoku Daily: History of Sudoku." http://www.sudokudaily.net/history.php.

Herzberg, Agnes, and Ram M. Murty. "Sudoku Squares and Chromatic Polynomials." *Notices of the American Mathematical Society* 54, no. 6 (2007).

Holly Hirst

Swimming

Category: Games, Sport, and Recreation.
Fields of Study: Data Analysis and Probability; Measurement; Number and Operations.
Summary: Swimming performance can be modeled and improved mathematically.

Mathematical modeling and statistical analysis have been applied to swimming in a variety of ways. Modeling the properties of fluids in motion is the subject of fluid dynamics, a sub-branch of mechanics. Placing objects in the fluid complicates the physics enormously.

The interaction of the fluid and object at the point where the object meets the fluid (called the "boundary") is of particular interest. Problems studied in this way include why flags "flap" in a breeze and how fish swim. Statistical analysis has been applied to a number of questions about swimming performance, including the prediction of future world record times, the modeling of deterioration in swimming performance as a function of age, and the evaluation of whether triathlons are fair to swimmers.

Improving Human Performance in Swimming

Modeling human swimming presents serious challenges for researchers. The use of arms and legs to propel the swimmer through the water adds complexity to the fluid dynamics models. Because the human swimmer is not completely immersed in the water but keeps part of the body above the surface, the interaction between the swimmer and the surface is particularly difficult to model. Researchers have applied smoothed particle hydrodynamics to the study of human swimming performance which, unlike traditional fluid dynamics, treats fluid flow as the motion of individual particles. This method enables researchers to more accurately model and simulate the interactions of the swimmer at the surface. The goal of this research is to help individual swimmers improve their performance in competition.

Predicting World Record Swim Times

Statistical analysis of human swimming performance encompasses a number of different approaches and methods. An analysis of world records in swimming from 1960 to 2010 shows a nearly steady decrease in times, resulting in between 15% to slightly more than 25% improvement, depending on the event. The question remains how long times can continue to decrease, how much is because of increased participation in swimming (especially women's swimming), and how much is because of advances in technique and conditioning.

Predicting the Swimming Performance of Aging Swimmers

On a different tack, Ray Fair modeled the performance of elite swimmers of different age groups and modeled the performance at various swimming distances by age. For example, he predicted that a 60-year-old will swim a time about 10% slower than the swimmer had done at age 35, while a 70-year-old will be 25% slower.

Are Triathlons Fair to Swimmers?

Richard De Veaux and H. Wainer investigated the relative disadvantage of swimmers to runners and cyclists in a triathlon. Because the times taken for the three events are so different, they argued that the standard triathlon proportions (including the Iron Man and Olympic triathlons) are grossly unfair to swimmers. The best marathon runners in the world take about two hours, seven minutes to run the 26.2-mile marathon (with variation due to course and weather). An elite cyclist can cover about 60 miles in the same time, and an elite swimmer can travel 7.5 miles. Thus, to be fair in terms of average time taken, a triathlon based on a marathon should also contain a 60-mile bike leg and a 7.5 mile swim. In reality, the Iron Man is a 26.2-mile run, a 112-mile bike leg, and only a 2.4-mile swim, and thus disadvantages swimmers enormously.

Further Reading

Cohen, R. C. Z., P. W. Cleary, and B. Mason. "Simulations of Human Swimming Using Smoothed Particle Hydrodynamics." In *Proceedings of the Seventh International Conference on CFD in the Minerals and Process Industries*. Melbourne, Australia: CSIRO, 2009.

———. "Improving Understanding of Human Swimming Using Smoothed Particle Hydrodynamics." *6th World Congress on Biomechanics* (2010).

Fair, Ray C. "How Fast Do Old Men Slow Down?" *Review of Economics and Statistics* 76, no. 1 (1994).

Wainer, H., and R. D. De Veaux. "Resizing Triathlons for Fairness." *Chance* 7 (1994).

———. "Making Triathlons Fair: The Ultimate Triathlon." *Swim Magazine* 10, no. 6 (1994).

Richard De Veaux

Symmetry

Category: Architecture and Engineering.
Fields of Study: Geometry; Measurement.
Summary: An ancient mathematical concept, there are various forms of symmetry.

"Symmetry," which comes from the Greek word roots meaning "same" and "measure," describes a picture, shape, or other object that looks the same when viewed from another perspective or that can be transformed in some way without changing its important properties. The word "symmetry" can refer to this property, to the transformation itself, or more holistically to an aesthetically pleasing sense of balance. Eighteenth-century mathematician Adrien-Marie Legendre revolutionized the concept of symmetry when he connected it to transformations. There are a wide variety of uses of the word "symmetry" in different domains, including art, architecture, and science, and many of these have existed from antiquity. The concept of symmetry is inherent to modern science and architecture, and its evolution reflects in many ways the dynamic nature of these fields.

Visual Symmetry

In the context of geometric figures drawn in the plane, there are three fundamental types of symmetry:

1. A figure has "reflection" symmetry if it coincides with its own mirror image across some line. The capital letters M and W have a single reflection symmetry, while the letter H has two symmetries, horizontal and vertical.
2. A figure has "rotational symmetry" if it can be rotated around a fixed point, leaving the figure unchanged. For example, the capital letters N, Z, and S are unchanged when rotated 180 degrees. The pattern of black squares in traditional crossword puzzles also has this half-turn symmetry.
3. A figure has "translational symmetry" if it can be slid or moved without changing. A typical example is a repeating pattern on wallpaper.

Construed in the broadest terms, symmetry plays a role in almost all art and is related to balance and harmony. One of the many ways in which the narrower geometric notion of symmetry applies to art is tessellations. A tessellation is a covering of the plane by copies of a limited set of tiles. Such figures are often highly symmetric. Tilings by squares, hexagons, and triangles are common enough, both in art and on kitchen floors, and more fanciful tessellations involving animal and plant shapes are also possible. Tessellations, dynamic symmetry, and mathematical sophistication are especially evident, for example, in the art of M.C. Escher (1898–1972).

Abstract Symmetries

Symmetry is not just a geometric concept. Any structure or object can have symmetry. Abstractly, a symmetry is any transformation of an object resulting in an object that is "the same" in the sense of having all the same properties that are important in context. Often, the object is a geometric figure, and the relevant properties are length, angle, and area, but it need not be so.

Consider the game rock-paper-scissors. Renaming the scissors gesture to "paper," renaming paper to "rock," and renaming rock to "scissors" would leave the rules of the game unaltered. This is an abstract symmetry of the game. Then, there are enough symmetries to identify any move with any other, so all three options are intrinsically "equally good." In this example, there is symmetry but no geometry whatsoever.

Symmetry and Groups

In higher mathematics, notions of symmetry are expressed in the language of group theory. A "group" is a set (G) of objects that can be composed together (in other words, if x and y are elements in a group, $x \times y$ is also an object in the group), subject to three conditions: associativity, identity, and inverse criteria. The salient feature of this definition is that the set of all the symmetries of any object satisfies these conditions. The associativity property is automatic from function composition; but what about the other two? These are restatements of the convention that the transformation that does nothing is a symmetry and the idea that symmetries are "undo-able." Symmetries leave an object "structurally the same as it was," so there will always be another symmetry to undo any given symmetry.

The symmetries of any object that preserve any desired features form a group, called the "symmetry group" of the object. Often, one can understand a complicated object much better by studying the size and structure of its symmetry group.

Klein and the Erlangen Program

Felix Klein (1849–1925) greatly strengthened the connection between geometry and group theory. His insight was that, if one really wants to understand a geometric structure, then one should study the group of symmetries that preserve the structure. This philosophy has proved very fruitful and is now known as the Erlangen program.

For example, in ordinary Euclidean plane geometry, the focus is on lengths and angles. The group of symmetries that preserve lengths and angles consists of translations, rotations, reflections, and combinations of these. Given any two points, each with an arrow pointing away from it in a given direction, one can always translate and rotate the plane so that the image of the first point lies on the second point, and the arrows are pointed in the same direction. This is the sophisticated way to understand the notion that every point and direction in the plane are functionally the same as every other point and direction.

The Erlangen program has played a fundamental role in the development of nineteenth- and twentieth-century geometric thinking, clarifying the relationships and distinctions between geometry and topology; projective and affine geometry; and Euclidean, hyperbolic, and spherical geometry.

Symmetry and the Universe

Those who study the shape of space are greatly concerned with symmetry. Consider the question of whether the universe is "homogeneous." That is, do the laws of physics treat every place the same as every other place? Is every direction physically like every other direction? What answers to those questions are believed to be correct determines what shapes, structures, and geometries are viable candidates to model the universe.

Time symmetry is another issue of importance in physics research; one wants to know to what extent the physical laws of the universe treat the past and future symmetrically. On a small enough scale, particle interactions have time symmetry. If one watched a "movie" of particle interactions on a small enough time-scale, it would be impossible to tell whether the movie was playing forward or backward. On the other hand, the large-scale events observed in everyday life do not possess such past-future symmetry; for example, eggshells break but do not spontaneously assemble, people age but do not become more youthful. This discrepancy between small-scale symmetry and large-scale asymmetry is rather mysterious, and one can hope that reconciling the two will lead to greater understanding of physics.

Symmetry and Architecture

Symmetry has long been connected with architecture. In Greek and Latin, symmetry was used to indicate a common measure or a notion of something well-proportioned, rather than as a reflection. However, reflection symmetries can be found in many buildings from different cultures, where the left side is a mirror image of the right side. Architects have also used symmetry in external views, layout, stability, or building details, such as stairs or windows. Some authors claim that the first recorded instance of the use of symmetry as a mirror reflection was in 1665, when Gian Lorenzo Bernini was asked to design an altar for the church of Val-de-Grace, while others assert that it was first found in Claude Perrault's 1673 treatise on columns. Perrault is best known as the architect of the east wing of the Louvre.

Concepts such as the symmetry groups of the plane also originate in architecture. Beginning with mathematician Edith Muller's 1944 analysis, experts continue to debate how many of the 17 groups can be found in the mosaics of the Alhambra at Granada, a fourteenth-century Moorish palace. Some assert that all 17 can be found there and in many other examples of Islamic art and architecture. A formal mathematical proof that there are no additional symmetry groups was proven independently by Evgraf Fedorov in 1891 and George Pólya in 1924. Partly because of a prohibition against using anthropomorphic forms, symmetry appears in many instances of Islamic-influenced architecture, such as the Taj Mahal.

The connections between symmetry and architecture continue into the twenty-first century. In numerous texts in the twentieth and twenty-first centuries, mathematicians such as Hermann Weyl illustrate concepts using architectural references. Architects and engineers also frequently use symmetry, though architects working in the modernist aesthetic reject symmetry in their designs.

Further Reading

Cohen, Preston Scott. *Contested Symmetries and Other Predicaments in Architecture*. Princeton, NJ: Princeton Architectural Press, 2001.

Gardner, Martin. *The Ambidextrous Universe: Symmetry and Asymmetry From Mirror Reflections to Superstrings*. 3rd ed. New York: W. H. Freeman, 1990.

Hon, Giora, and Bernard Goldstein. *From Summetria to Symmetry: The Making of a Revolutionary Scientific Concept*. New York: Springer, 2008.

Weyl, Hermann. *Symmetry*. Princeton, NJ: Princeton University Press, 1952.

Michael "Cap" Khoury

Textiles

Category: Arts, Music, and Entertainment.
Fields of Study: Geometry; Representations.
Summary: Mathematics is integral to creating both traditional and modern weave patterns in textiles.

Textiles are flexible sheets made out of fibers. Natural textiles are made from plants, animals, or minerals; artificial textiles use human-made fibers, like plastic or synthetic proteins. Woven textiles combine longer fiber threads either by hand or by using looms or knitting machines. In nonwoven textiles, like felt, short or microscopic fibers are bonded by chemical or physical treatments. Nonwovens are often meant to be highly durable or disposable and have many applications in health, construction, and filtration technologies. Mathematical methods are used to design, produce, and analyze textiles. In 1804, Joseph Jacquard invented a weaving system using cards with patterns of holes to control loom threads. These cards were later modified by Charles Babbage into computer punch cards. Weaver and mathematics teacher Ada Dietz wrote *Algebraic Expressions in Handwoven Textiles* in 1949. She outlined a method for using expansions of multivariate polynomials to generate weaving patterns.

Weave Formulas

On a loom, "warp" threads are held parallel and "weft" threads are passed over and under them. A pattern formed in one pass of weft can be either repeated exactly, transposed, or otherwise changed in the next passes. Let A stand for warp threads on top and E stand for the weft thread on top. In plain fabric, a pattern $AEAE\ldots$ indicates that the weave is transposed by one thread in the next row. Basket weave uses $AAEEAAEE\ldots$, so the pattern is repeated for two rows and then transposed by two (or some other whole number) for the next two rows. Satin is $AAAAEAAAAE\ldots$, giving four repeats followed by one transposition. A satin weave results in the majority of the threads being parallel, so light is minimally scattered, producing the characteristic sheen. In contrast, twill has a distinct, textured diagonal pattern formed by using an $EEAEEA\ldots$ weaving scheme. Patterns may be added to plain weaves by printing or dying the fabric. The U.S. group Complex Weavers provides a forum for sharing advanced weaving methods and patterns, such as manifold twill.

Patterns and other factors like the thread intersections per area also dictate other properties. For example, plain weave fabrics tear the easiest, because force is applied to the single thread immediately next to the tear. Crimp is how easily the fabric morphs under tension. Plain weaves generally morph the easiest. Wrinkle resistance is the opposite; the more freedom of movement threads have, the easier it is for them to return to smoothness. Satin is an example of a wrinkle-resistant weave. On the other hand, satin silks shrink the most because their weave pattern is loose. Twill has a relatively high resistance to tearing, which makes fabrics such as jean popular for working clothing.

Cultural Textiles

Textiles are a significant cultural art form for many people in Africa. The three most well-known forms are *kente*, *adire*, and *adinkra*. Kente cloth is woven in long narrow strips, traditionally by Asante and Ewe men, and then sewn together into larger pieces of fabric that may be used for clothing or household goods. The cloth was often a sign of wealth and kept for special occasions. There are more than 300 known kente patterns, many of which represent people or historical events. Widely found adira cloth has patterns made by resistance dying. The cloth is tied, stitched, or stenciled, often with geometric patterns, to prevent the dye from adhering to some portions of the cloth. Adinkra cloth is printed, usually by drawing a square grid and stamping symbols into each square. This highly developed symbol language expresses concrete and abstract concepts, such as transformation or unity. Like kente cloth, adinkra often tells stories or proverbs. Tessellations and other repeating patterns are also common. In Ghana, the cloth was originally worn for mourning and some is still reserved for that purpose.

In Scotland, tartans represent families, clans, or regions. A "sett" is a specific plaid pattern, specified by sequences and widths of colored stripes. The pattern is formed by interweaving bands of stripes at right angles. Most are symmetrical, which means the sett is reflected

90 degrees around a pivot or center stripe. Asymmetrical setts have no pivot point. Symmetry has implications in kilt making. A kilt "pleated to the sett" has pleats folded to visually reproduce the tartan pattern across the back of the kilt, often not possible with an asymmetric pattern. Tartan patterns have been investigated with mathematical methods, such as group theory, and they are used in classrooms as examples of symmetry. Artist Andrew Hennessey has proposed "stella tartan" in which tartan setts would be woven radially and overlap in irregular polygon patterns.

High-Technology Textiles

The Industrial Revolution made rapid mass production of textiles feasible and the textile industry has since used many mathematical and computational techniques to continue its evolution. These techniques include differential equations, numerical methods, image processing, pattern recognition, and statistics. Computer-aided design (CAD) and computer-aided looms (CAL) are widespread. Application areas include supply chain management, quality control, and product development. The latter may involve structural modeling and simulation, as well as thermal or biomechanical bioengineering, particularly for specialty textiles. Some competitive swimwear has tiny triangular projections that mimick shark skin to reduce drag. An absorbent, nonwoven textile called *air-laid paper* is used in diapers. Integrating tiny light-emitting diodes into fabric allows clothes to change color or display text or animation. Thermal self-regulation may be achieved with phase-changing microcapsules that become fluid for cooling or solid to release heat, as needed. Weak link theory and bundle theory, as well as research in twisted continuous filaments, helical modeling of yarns, two-dimensional elasticity theory, aerodynamics, and many other investigations have also revolutionized the individual threads that compose fabric, often changing its properties even when using traditional weaves.

Further Reading

Dietz, Ada. *Algebraic Expressions in Handwoven Textiles.* Louisville, KY: The Little Loomhouse, 1949. http://www.cs.arizona.edu/patterns/weaving/monographs/dak_alge.pdf.

Harris, Mary. *Common Thread: Women, Mathematics and Work.* Staffordshire, England: Trentham Books Limited, 1997.

Zeng, Xianyi, Yi Li, Da Ruan, and Ludovic Koehl. *Computational Textile.* Berlin: Springer, 2007.

Maria Droujkova

Tic-Tac-Toe

Category: Games, Sport, and Recreation.
Fields of Study: Geometry; Number and Operations; Problem Solving.
Summary: Traditional Tic-Tac-Toe has a limited number of possible games, which can lead players to quickly discover an unbeatable strategy as long as they move first.

Tic-tac-toe is a famous game often played by children. It requires a playing board of a 3-by-3 arrangement of square cells, usually quickly drawn by making two vertical lines cross two horizontal lines and imagining an outer border. Two players alternate marking cells with either an X (usually the first player) or an O (the second player). Each attempts to put three of their marks in a straight line, while trying to block the attempts of the other. The winner is the player who first makes the three-in-a-row line. Unfortunately for the challenge of the game, the first player can always win by putting an X in the center cell and playing carefully. Children often learn this strategy and the game can become mundane if this strategy is always employed.

Play Possibilities

However, tic-tac-toe is simple enough that it can serve as a fairly easy example of game analysis, where all possible positions and plays are determined. Most other games are so complex that such analyses are overwhelmingly complex.

Ignoring symmetric patterns, there are three possible first plays—a corner, a side, or the center. The second play patterns are based on these three openings. Again, ignoring symmetries, the corner opening leads to five possible second moves, the side opening also allows five possible second moves, but following a center opening there are only two possible second plays. Hence, there are a total of 12 noncongruent, nonsymmetrical second plays. Similar exploration of the possibilities shows a total of 66 possible third moves, though

26 are duplications, so there are only 40 noncongruent arrangements after the third play. Then, it becomes much more complicated because of overlaps of first- and third-move Xs and second- and fourth-move Os. This fact demonstrates that even in such a simple game as tic-tac-toe, the full analysis becomes quite complex.

Variations

The 3-by-3 magic square (with numbers 1–9 arranged in the cells so that each row, column, and diagonal sums to 15) looks like a tic-tac-toe board with numbers. A game can be played where players take turns choosing numbers 1–9 (without repeats), trying to reach a sum of 15 with three numbers. Playing this game and placing the numbers onto the 3-by-3 magic square turns out to follow the same general games strategies as tic-tac-toe.

Tic-tac-toe can become a much more interesting—and challenging—game by expanding the board to three dimensions. If the game is played on a stack of three 3-by-3 boards (a cube of 27 cells), any row of three is a win. Some have suggested that a 4-by-4-by-4 cube, with a line of four to win, is a smoother game. Winning lines can lie entirely on a horizontal level, drop vertically from top to bottom, slant along a vertical plane, or go from one corner to the opposite corner along the body diagonal. New players often have difficulty even noticing winning lines! For even more complexity, the game can be played in four dimensions, usually displayed as a two-dimensional array of two-dimensional boards, assuming the boards can be stacked in any of the horizontal, vertical, or diagonal ways, with winning lines in any of the stacks according to the three-dimensional patterns, a variation that can be either 3-by-3-by-3-by-3 or 4-by-4-by-4-by-4.

Alternatively, the traditional board can be imaged to extend infinitely, allowing more possibilities for winning lines. One version keeps the traditional board but assumes the left column wraps to be next to the right column, so a line of three can be the upper center, the right center, and the left bottom corner. Similarly, the top and bottom rows can be considered as wrapping around to be next to each other.

Nine-Men's Morris

Many games from around the world pick up on the ideas of tic-tac-toe, especially the goal of making three (or more) counters in a row. Probably the most famous is called Nine-Men's Morris in English (also called "mill" or, in French, *merelles* or *morelles*); some suggest early versions were even played in ancient Egypt. The board is three concentric squares connected in the middles of the sides, with each junction and corner marked with a dot. Two players each have nine counters, marked to distinguish those of each player. They take turns playing their counters onto the dots of the board, trying to get three in a row, which is called a "mill." After players use up the nine counters each, play continues by sliding already-played counters along the lines on the board. Anytime a row of three is made by one player, the player is allowed to remove one of the other player's counters (but they cannot take a counter that is already in a mill). Eventually, one player either has no counters left or cannot move any remaining counters, and the other player wins.

Further Reading

Beck, Jozsef. "Combinatorial Games: Tic-Tac-Toe Theory." In *Mathematical Constants: Encyclopedia of Mathematics and its Applications*. Edited by Steven R. Finch. New York: Cambridge University Press, 2008.

Malumphy, Chris. "3-D Tic-Tac-Toe." http://home.earthlink.net/~cmalumphy/3d.html.

Masters, James. "Nine Mens Morris, Mill—Online Guide." http://www.tradgames.org.uk/games/Nine-Mens-Morris.htm.

Smit, William. "4-D Tic-Tac-Toe Game." http://www.ugcs.caltech.edu/~willsmit/4d/index.html.

Zaslavsky, Claudia. *Tic Tac Toe: And Other Three-In-A Row Games From Ancient Egypt to the Modern Computer*. Toronto: Crowell, 1982.

Lawrence H. Shirley

Volleyball

Category: Games, Sport, and Recreation.
Fields of Study: Data Analysis and Probability; Geometry; Measurement.
Summary: Mathematics is fundamental to player motion, strategy, and scoring in volleyball.

Volleyball, which began in the late nineteenth century as a non-contact recreational sport, quickly developed into a globally popular competitive sport. Two teams, typically with two to six players, face one another on opposite sides of a rectangular court divided by a net. Beach volleyball is played on sand courts rather than a hard surface. Game strategy uses mathematical concepts such as angles, rotation, and parabolic motion in an effort to impart optimal trajectories, speeds, and spins on the ball to prevent the other team from successfully returning it. The receiving team must understand three-dimensional motion and vectors in order to intercept the ball and change its direction, often using a sequence of hits coordinated among several players. The strategies of beach volleyballers often differ from those of hard court volleyballers because of differences in the ability to jump or dive for an incoming ball. Mathematics is also used to analyze and model body kinetics, such as the motions of a player's shoulders and arms while serving. Statistics are used to analyze and describe both team and individual proficiencies and success. These include measures like number of attacks, kills, and assists; hitting percentages; and kill average and efficiency as a function of total attempts.

General Game Play and Scoring

Volleyball teams work together to hit the ball over the net in such a way as to prevent the other team from returning it. A match consists of three or five games. The third game of a three-game match or the fifth game of a five-game match is the deciding game. A single sequence of back and forth hitting is known as a "rally," which begins with one side serving the ball and ends when one team or the other fails to legally return it. Each side gets three attempts and the same player may not touch the ball twice in a row. At the end of the rally, the winning side may earn a point, the right to serve the ball, or both.

There are two different scoring systems used in volleyball. In side-out scoring, only the serving team may earn a point. In rally scoring, either side earns a point. Winners always get the serve. Deciding games are played to 15 points; nondeciding games are played to 25. However, the winning team must be ahead by at least two points or play continues. Sometimes, a scoring cap is used, which nullifies this requirement. Statistical analyses show that rally point scoring makes matches shorter and match lengths more predictable versus side-out scoring. However, there appears to be no significant effect on scoring margins between teams; on average, after an even number of serve changes, points awarded to non-serving teams balance. In addition to statistics, Markov chains are useful for analyzing volleyball games in terms of the proportion of points won and the probabilities of winning a point, game, and match.

Player Roles and Strategy

Hard court teams typically consist of six players with specialized roles, with the left, center, and right forwards in a row along the frontcourt and the left, center, and right backs in a row along the backcourt. However, players usually rotate through positions during play, requiring analysis of permutations and the timing of substitutions. Beach volleyball teams typically consist of two players each, generally front and back. Players seek to control the ball through the angle, force, and timing with which the ball is struck and by choosing whether or not to impart spin on the ball. The volleyball typically travels along a parabolic path, modified by its spin and additionally influenced by player efforts and external factors, such as air resistance. The basic skills used in volleyball include the serve, pass, set, spike, block, and dig. A variety of serves can be used as the server hits the ball into the opponent's court. Different types of serves affect the ball's direction, speed, and acceleration with the goal of increasing the difficulty of handling the ball for the opposing team. Serves that have flatter parabolic paths tend to preserve more of the initial force and velocity and are usually more difficult to return.

The opposing team's first reception of the ball is known as the "pass," the second contact is known as the "set," and the third contact is known as the "attack" (also called "spike"), though a team may not opt to use all three contacts in every play. A block is a team's attempt to prevent the opposite team from spiking the ball into their court, and a dig is an attempt to prevent a ball from hitting the court. Shots include the hard angle, deep angle, seam shot, line shot, angled line shot, swiping shot, high and hard, and the save. Achieving different shots relies on affecting the ball's speed, spin, and angle of trajectory through shoulder and hip positions, aiming at gaps between opposing players, and the amount of force applied. Spin tends to make the ball more difficult to return successfully, since the appropriate counterforce to control the ball and change its directional vector is more difficult to determine and apply quickly.

Further Reading

Calhoun, William, G. R. Dargahi-Noubary, and Yixun Shi. "Volleyball Scoring Systems." *Mathematics and Computer Education* (Winter 2002).

Kiernan, Denise. *Sports Math*. New York: Scholastic Professional, 1999.

USA Volleyball. *Volleyball Systems and Strategies*. Champaign, IL: Human Kinetics, 2009.

Marcella Bush Trevino

Wind Instruments

Category: Arts, Music, and Entertainment.
Fields of Study: Geometry; Number and Operations; Representations.
Summary: The frequency and pitch of wind instruments are determined by their shape, length, and other factors.

Wind instruments convert the energy of moving air into sound energy—vibrations that are perceptible to the human ear. Under this definition, wind instruments include the human voice; pipe organs; woodwind instruments, such as the clarinet, oboe, and flute; and brass instruments, like the trumpet. The nature of this vibration and the associated resonator tube are responsible for the unique timbre of each type of wind instrument.

Sources of Vibrations

In the human voice, the flow of air from the lungs causes the vocal cords (also called "vocal folds") in the larynx to open and close in rapid vibration. This periodic stopping of the air stream creates oscillatory pulses of air pressure, or sound. The frequency of this vibration and the pitch of the resulting sound are determined by the length and tension of the cords. A singer or speaker controls these factors using the musculature of the larynx.

The rapid open-close vibration of the vocal cords is present in many wind instruments. In brass instruments, such as the trumpet, trombone, French horn, and tuba, the lips of the musician form a small aperture that opens and closes in response to air pressure. Brass instruments are sometimes called "lip-reed" instruments. In single-reed instruments, like the clarinet and saxophone, a thin cane reed vibrates in oscillatory contact with a specially shaped structure (the mouthpiece) to bring about the open-close effect. The oboe and bassoon utilize two cane reeds held closely together with a small space between them that opens and closes in response to flowing air, controlled by the muscles of the lips.

A third important mechanism for converting the energy of moving air into vibration is utilized in the flute and the so-called flue pipes of the pipe organ. In these instruments, vibration occurs when flowing air passes over an object with a distinct edge that splits the airstream. The resulting turbulence gives rise to oscillatory vibration. With the modern flute, the flutist's lip muscles actively control the interaction between the airstream and the edge. With the recorder and other whistle-type instruments, as well as flue pipes of the organ, the interaction is controlled by the mechanical design of the instrument alone.

Tube Resonators and Overtones

With the exception of the human voice, all wind instruments are constructed with a tube resonator enclosing a column of air that functions in much the same way as the vibrating string. Oscillations in air pressure inside the tube reflect from the ends, resulting in significant feedback with the primary vibrating medium. The relationship between the vibration frequency and length of a string fixed at both ends is explained by the concept of "harmonics." In idealized settings, changing the string length by small integer factors (for example, 1/2, 1/3, or 1/4) results in frequency changes that are recognizable as musical intervals (for example, an octave, an octave plus a fifth, or two octaves). The resonating air column in wind instruments behaves similarly to a vibrating string.

(Photos.com)

An important performance practice on most wind instruments is overblowing. Not to be confused with simply playing overly loudly, the term "overblowing" refers to the fact that changes in the airflow can cause the resonating air column to vibrate at an overtone above its fundamental frequency. Overblowing allows performers on modern instruments to achieve a large range of pitches (often two octaves or more) from a relatively compact resonating tube. Instruments with cylindrical tubes open at both ends, such as in some flutes, overblow at the octave, as do conical instruments that are closed at one end, such as the oboe and saxophone. On the other hand, cylindrical tubes closed at one end, such as the clarinet, overblow at the twelfth—an octave plus a fifth. The relative weakness of the overtone at the octave and other even-numbered overtones account for the particular timbre of the clarinet.

Altering the Tube Length in Performance

Just as the length of a vibrating string determines the frequency or pitch of the vibration, the length of the resonating air column accounts for the pitch of notes played by a wind instrument. In reed instruments, the resonating tube is perforated along its length with holes. By systematically covering some of the holes but not others, the musician effectively changes the length of the resonating column. This change, in turn, causes the vibrating reed assembly to assume the frequency of the air column. Most brass instruments have secondary lengths of tubing that are brought into play by mechanical valves by which the performer alters the length and the fundamental frequency of the vibrating air column. The exception to this is the slide trombone, which features a concentric tube arrangement by which the outer tube can move to lengthen the air column resonator.

Further Reading

da Silva, Andrey Ricardo. *Aeroacoustics of Wind Instruments: Investigations and Numerical Methods*. Saarbrücken, Germany: VDM Verlag, 2009.

Miller, Dayton Clarence. *The Science of Musical Sounds*. Charleston, SC: Nabu Press, 2010.

Sundberg, Johan. *The Science of Musical Sounds: (Cognition and Perception)*. San Diego, CA: Academic Press, 1991.

Wood, Alexander. *The Physics of Music*. 7th ed. London: Chapman and Hall, 1975.

Eric Barth

Resource Guide

Books

Aaboe, Asger. *Episodes From the Early History of Mathematics*. Washington, DC: Mathematical Association of America, 1975.

Adrian, Yeo. *The Pleasures of Pi and Other Interesting Numbers*. Singapore: World Scientific Publishing, 2006.

Agresti, A. *Categorical Data Analysis*. Hoboken, NJ: Wiley, 2002.

Aho, A. V., J. E. Hopcrotf, and J. D. Ullman. *The Design and Analysis of Computer Algorithms*. Reading, MA: Addison-Wesley, 1976.

Albert, Jim, and Jay Bennett. *Curve Ball: Baseball, Statistics, and the Role of Chance in the Game*. New York: Springer-Verlag, 2001.

Ascher, Marcia. *Mathematics Is Everywhere: An Exploration of Ideas Across Cultures*. Princeton, NJ: Princeton University Press, 2002.

Ball, W. W. Rouse. *A Short Account of the History of Mathematics*. New York: Sterling Publishing Company, 2001.

Barnett, Raymond, Michael Ziegler, and Karl Byleen. *Calculus for Business, Economics, Life Science, and Social Science*. Upper Saddle River, NJ: Prentice-Hall, 2005.

Baumohl, Bernard. *The Secrets of Economic Indicators: Hidden Clues to Future Economic Trends and Investment Opportunities*. 2nd ed. Upper Saddle River, NJ: Pearson Education, 2008.

Beckmann, Petr. *A History of π (Pi)*. New York: Barnes & Noble, 1971.

Behrends, Ehrhard. *Five-Minute Mathematics*. Providence, RI: American Mathematical Society, 2008.

Bell, Eric Temple. *Men of Mathematics*. New York: Simon & Schuster, 1937.

Bennett, Jay, and James Cochran. *Anthology of Statistics in Sports*. Philadelphia, PA: Society for Industrial and Applied Mathematics, 2005.

Berggren, Lennart, Jon Borwein, and Peter Borwein. *Pi: A Source Book*. New York: Springer-Verlag, 1997.

Berlekamp, Elwyn R., John H. Conway, and Richard K. Guy. *Winning Ways for Your Mathematical Plays*. Natick, MA: AK Peters, 2001.

Blackwell, William. *Geometry in Architecture*. Hoboken, NJ: Wiley, 1984.

Blatner, David. *The Joy of π*. New York: Walker & Co., 1997.

Blue, Ron, and Jeremy White. *The New Master Your Money: A Step-by-Step Plan for Gaining and Enjoying Financial Freedom*. Chicago: Moody, 2004.

Blum, Raymond. *Mathemagic*. New York: Sterling Publishing, 1992.

Bodie, Zvi, Alex Kane, and Alan Marcus. *Investments*. Chicago, IL: McGraw-Hill/Irwin, 2008.

Borwein, Jonathan, and Peter Borwein. *A Dictionary of Real Numbers*. Pacific Grove, CA: Brooks/Cole Publishing Co., 1990.

Boyer, C. B. *A History of Mathematics*. Hoboken, NJ: Wiley, 1968.

Boyer, C. B. *The History of the Calculus and Its Conceptual Development*. New York: Dover Publications, 1949.

Brealey, Richard A., Stewart C. Myers, and Franklin Allen. *Principles of Corporate Finance*. 9th ed. New York: McGraw-Hill, 2008.

Bressoud, David. *The Queen of the Sciences: A History of Mathematics*. Chantilly, VA: The

Teaching Company, 2008.

Broverman, Samuel A. *Mathematics of Investment and Credit*. Winsted, CT: ACTEX Publications, 2008.

Burkett, Larry, and Brenda Armstrong. *Making Ends Meet: Budgeting Made Easy*. Gainesville, GA: Crown Financial Ministries, 2004.

Burton, David M. *The History of Mathematics: An Introduction*. New York: McGraw-Hill, 2005.

Calinger, Ronald. *A Contextual History of Mathematics*. Upper Saddle River, NJ: Prentice-Hall, 1999.

Clagett, Marshall. *Archimedes in the Middle Ages*. Madison: University of Wisconsin Press, 1964.

Closs, Michael. *A Survey of Mathematics Development in the New World*. Ottawa: University of Ottawa, 1977.

Closs, Michael, ed. *Native American Mathematics*. Austin: University of Texas Press, 1986.

Coe, Michael D. *Breaking the Maya Code*. New York: Thames and Hudson, 1992.

Copeland, Thomas E., J. Fred Weston, and Kuldeep Shastri. *Financial Theory and Corporate Policy*. 4th ed. Upper Saddle River, NJ: Pearson Education, 2005.

Cullen, Christopher. *Astronomy and Mathematics in Ancient China: The Zhou Bi Suan Jing*. Cambridge, England: Cambridge University Press, 1996.

Cuomo, Serafina. *Ancient Mathematics*. London: Routledge, 2001.

Davenport, Harold. *The Higher Arithmetic: An Introduction to the Theory of Numbers*. Cambridge, England: Cambridge University Press, 1999.

Davis, Morton D. *The Math of Money: Making Mathematical Sense of Your Personal Finances*. New York: Copernicus, 2001.

De Mestre, Neville. *The Mathematics of Projectiles in Sport*. Cambridge, England: Cambridge University Press, 1990.

Devlin, Keith. *The Math Gene: How Mathematical Thinking Evolved and Why Numbers Are Like Gossip*. New York: Basic Books, 2001.

———. *The Unfinished Game: Pascal, Fermat, and the Seventeenth-Century Letter That Made the World Modern*. New York: Basic Books, 2008.

Drobat, Stefan. *Real Numbers*. Upper Saddle River, NJ: Prentice-Hall, 1964.

Dudley, Underwood. *Numerology or What Pythagoras Wrought*. Washington, DC: Mathematical Association of America, 1997.

Eastway, Rob, and John Haigh. *Beating the Odds: The Hidden Mathematics of Sport*. London: Robson Books, 2007.

Eglash, Ron. *African Fractals: Modern Computing and Indigenous Design*. New Brunswick, NJ: Rutgers University Press, 1999.

Eves, Howard. *An Introduction to the History of Mathematics*. New York: Saunders College Publishing, 1990.

Flegg, G. *Numbers: Their History and Meaning*. New York: Schocken Books, 1983.

Friberg, Jöran. *Unexpected Links Between Egyptian and Babylonian Mathematics*. Singapore: World Scientific Publishing Co., 2005.

Friedman, Arthur. *World of Sports Statistics: How the Fans and Professionals Record, Compile, and Use Information*. New York: Athenaeum, 1978.

Fries, Christian. *Mathematical Finance: Theory, Modeling, Implementation*. Hoboken, NJ: Wiley, 2007.

Frumkin, Norman. *Guide to Economic Indicators*. Armonk, NY: M. E Sharpe, 2000.

Gamow, George. *One, Two, Three... Infinity*. New York: Viking Press, 1947.

Gardner, David, and Tom Gardner. *The Motley Fool Personal Finance Workbook: A Foolproof Guide to Organizing Your Cash and Building Wealth*. New York: Fireside Books, 2003.

Gardner, Martin. *Mathematics, Magic and Mystery*. New York: Dover, 1956.

Gay, Timothy. *The Physics of Football*. New York: HarperCollins, 2005.

Gerdes, Paulus. *Geometry From Africa: Mathematical and Educational Explorations*. Washington, DC: Mathematical Association of America, 1999.

Gillings, R. J. *Mathematics in the Time of the Pharaohs*. New York: Dover Publications, 1982.

Gutstein, Eric, and Bob Peterson, eds. *Rethinking Mathematics: Teaching Social Justice by the Numbers*. Milwaukee, WI: Rethinking Schools, 2005.

Hadamard, Jacques. *A Mathematician's Mind*. Princeton, NJ: Princeton University Press, 1996.

Hardy, G. H. *A Mathematician's Apology*. Cambridge, England: Cambridge University Press, 1941.

Henry, Granville C. *Logos: Mathematics and Christian Theology*. Lewisburg, PA: Bucknell University Press, 1976.

Hersh, Rueben. *What Is Mathematics, Really?* New York: Oxford University Press, 1997.

Hoyle, Joe Ben, Thomas F. Schaefer, and Timothy S. Doupnik. *Fundamentals of Advanced Accounting*. New York: McGraw-Hill, 2010.

Kalbfleisch, John D., and Ross L. Prentice. *The Statistical Analysis of Failure Time Data*. Hoboken, NJ: Wiley, 2002.

Katz, Victor J., ed. *Mathematics of Egypt, Mesopotamia, China, India, and Islam: A Sourcebook*. Princeton, NJ: Princeton University Press, 2007.

Kellison, Stephen G. *Theory of Interest*. New York: McGraw-Hill, 2009.

Kimmel, Paul D., Jerry J. Weygandt, and Donald E. Keiso. *Financial Accounting: Tools for Business Decision Making*. Hoboken, NJ: Wiley, 2009.

King, Jerry. *The Art of Mathematics*. New York: Plenum Press, 1992.

Klein, John P., and Melvin L. Moeschberger. *Survival Analysis: Techniques for Censored and Truncated Data*. New York: Springer-Verlag, 1997.

Kline, M., *Mathematical Thought From Ancient to Modern Times*. New York: Oxford University Press, 1972.

Koetsier, T., and L. Bergmans, eds. *Mathematics and the Divine: A Historical Study*. Amsterdam: Elsevier, 2005.

Longe, Bob. *The Magical Math Book*. New York: Sterling Publishing, 1997.

Martzloff, Jean-Claude. *A History of Chinese Mathematics*. New York: Springer-Verlag, 1987.

Moses, Robert P., and Charles E. Cobb, Jr. *Radical Equations: Civil Rights From Mississippi to the Algebra Project*. Boston: Beacon Press, 2001.

Mullis, Darrell, and Judith Handler Orloff. *The Accounting Game: Basic Accounting Fresh From the Lemonade Stand*. Naperville, IL: Sourcebooks, 2008.

Nahin, Paul J. *Dr. Euler's Fabulous Formula*. Princeton, NJ: Princeton University Press, 2006.

Nasar, Sylvia. *A Beautiful Mind: The Life of Mathematical Genius and Nobel Laureate John Nash*. New York: Simon & Schuster, 2001.

Oliver, Dean. *Basketball on Paper: Rules and Tools for Performance Analysis*. Washington, DC: Brassey's, 2004.

Pullan, J. M. *The History of the Abacus*. New York: F. A. Praeger, 1969.

Rafiquzzaman, M. *Fundamentals of Digital Logic and Microcomputer Design*. Hoboken, NJ: Wiley, 2005.

Rudin, W. *Principles of Mathematical Analysis*. New York: McGraw-Hill, 1953.

Salem, Lionel, Frédéric Testard, and Coralie Salem. *The Most Beautiful Mathematical Formulas*. Hoboken, NJ: Wiley, 1992.

Schwarz, Alan. *The Numbers Game: Baseball's Lifelong Fascination with Statistics*. New York: St. Martin's Press, 2004.

Smith, D. E. *History of Mathematics*. Vol. 2. New York: Dover Publications, 1958.

Solow, Daniel. *How to Read and Do Proofs: An Introduction to Mathematical Thought Process*. Hoboken, NJ: Wiley, 1982.

Steen, Lynn A. *On the Shoulders of Giants: New Approaches to Numeracy*. Washington, DC: National Academy Press, 1990.

Sterrett, Andrew. *101 Careers in Mathematics*. Washington, DC: The Mathematical Association of America, 1996.

Suzuki, Jeff. *A History of Mathematics*. Upper Saddle River, NJ: Prentice Hall, 2002.

Taylor, Alan D. *Mathematics and Politics: Strategy, Voting Power, and Proof*. New York: Springer-Verlag, 1995.

van der Waerden, B. L. *Geometry and Algebra in Ancient Civilizations*. Berlin: Springer, 1983.

Venema, G.A. *The Foundations of Geometry*. Upper Saddle River, NJ: Pearson Prentice Hall, 2006.

Weygandt, Jerry J., Paul D. Kimmel, and Donald E. Keiso. *Managerial Accounting: Tools for Business Decision Making*. Hoboken, NJ: Wiley, 2008.

Winkler, Peter. *Mathematical Puzzles: A Connoisseur's Collection*. Natick, MA: AK Peters, 2004.

Wright, Tommy, and Joyce Farmer. *A Bibliography of Selected Statistical Methods and Development Related to Census 2000*. Washington, DC: U.S. Bureau of the Census, 2000.

Yeldham, F. A. *The Teaching of Arithmetic Through Four Hundred Years (1535–1935)*. London: G. G. Harrap & Company, 1935.

Yong, L. L., and A. T. Se. *Fleeting Footsteps*. Singapore: Word Scientific Publications, 2004.

Zaslavsky, Claudia. *Africa Counts: Number and Pattern in African Culture*. Chicago: Lawrence Hill Books, 1999.

Zill, D. G. *Calculus with Analytic Geometry*. Boston: Prindle, Weber & Schmidt, 1985.

Journals and Magazines

The AMATYC Review

The American Mathematical Monthly

Association for Women in Mathematics Newsletter

Biometrics
Chance
The College Mathematics Journal
Experimental Mathematics
The Fibonacci Quarterly
Historia Mathematica
IMU-Net
Involve
Journal of Humanistic Mathematics
Journal of Integer Sequences
Journal of Recreational Mathematics
Journal of Statistics Education
Loci
MAA FOCUS
Math Horizons
Mathematics Magazine
Mathematics Teacher
NAM Newsletter
Notices of the American Mathematics Society
The Pentagon
Pi Mu Epsilon Journal
Plus Magazine
PRIMUS
Rose-Hulman Undergraduate Mathematics Journal
SIAM Review
Scholastic Math
Significance
Teaching Children Mathematics
Undergraduate Mathematics and Its Applications

Internet

American Institute of Mathematics
    www.aimath.org
The Algebra Project
    www.algebra.org
AMATYC
    www.amatyc.org
American Mathematical Society
    www.ams.org
American Statistical Association
    www.amstat.org
Association for Women in Mathematics
    www.awm-math.org
CryptoKids
    www.nsa.gov/kids
Datamath Calculator Museum
    www.datamath.org
Illuminations
    illuminations.nctm.org
MacTutor History of Mathematics
    www-history.mcs.st-and.ac.uk
Mathematical Fiction
    http://kasmana.people.cofc.edu/MATHFICT
Math for America
    www.mathforamerica.org
Math Forum
    www.mathforum.com
Math Fun Facts!
    www.math.hmc.edu/funfacts
MathDL
    mathdl.maa.org/mathDL
Mathematical Association of America
    www.maa.org
Mathematical Science Research Institute
    www.msri.org
The Museum of Mathematics
    www.momath.org
National Association of Mathematicians
    www.nam-math.org
National Council of Teachers of Mathematics
    www.nctm.org
RadicalMath
    www.radicalmath.org
Society for Industrial and Applied Mathematics
    www.siam.org
We Use Math
    www.weusemath.org
Wolfram MathWorld
    www.mathworld.wolfram.com

Index

Text and page numbers in **boldface** refer to main topics.

Abu al-Wafa Buzjani
    Those Parts of Geometry Needed by Craftsmen, 89
acrostics, word squares, and crosswords, 1–3
Adams, John (composer)
    Dr. Atomic, 98
Advanced Placement Calculus (AP Calculus), 30
Agon, 24
Alberti, Leone Battista, 102
album cover art, 106
Alexander, James Waddell, II, 73
Alexander Polynomial, 73
algebra and algebra education
    as core subject, 6
Algebraic Expressions in Handwoven Textiles (Dietz), 143
algebra in society, 3–9
    applications, 4, 6
    early history of, 4
    as gateway, 3, 7
    usefulness, 3, 6
algorithms
    Stable Matching, 80
Alice in Wonderland (Carroll), 21
"All is Number" (Pythagoras maxim), 98
American Crossword Puzzle Tournament, 2
American Mathematical Society (AMS), 38
animation and CGI
    principles of, 60
anthropometry, 33
APBR metrics, 15

ape index, 33
Arabic/Islamic mathematics
    religious tenets and, 88, 99
archery, 9–10
architecture
    careers in, 62
    geometry and, 57, 59
    golden ratio in, 66
    symmetry and, 142
    tessellations in, 89
arithmetic magic, 76
arithmetic puzzles, 110
Ark of the Covenant, 86
Ashton, Frederick
    Scènes de Ballet, 11
Asimov, Isaac
    The Realm of Algebra, 3
Associated Press (AP), 41
Association for Professional Basketball Research, 15
atonal music, 36
audio processing, 107
Augustine, Saint (Aurelius Augustinius), 86
automobiles
    design of, 26
    speedometers, 41
Auto-Tune software, 107
axiomatic systems, 57

Babbage, Charles
    weaving systems and, 143
Babylonian mathematics
    and religion, 98
Bach, Johann Sebastian, 36

Bacon, Kevin, 128, 129
Bacon, Roger, 86
Balan, Radu, 34
Ballanchine, George, 37
ballet, 10–11, 37
ballroom dancing, 11–13
Barrow, John, 86
base-10 system, 98
baseball, 13–15, 51
Baseball Abstract (James), 13, 51
basketball, 15–17
basketry, 17–18
battement tendu (ballet), 10
Baudot, Emile, 106
Baumé scale, 44
Bayes, Thomas, 100
Beaujoyeulx, Balthazar
    Le Balet Comique de la Reine, 11
bells, 105
Bells, 105
Benesh Movement Notation, 11
Benjamin, Arthur, 78
Bergman, C. A., 10
Berkeley, George, 25
Bernoulli-Euler equation, 10
betting and fairness, 19–20
Bhagavad Gita, 100
bianzhong bells, 105
bible codes, 100
billiards, 21–22
Bill James Baseball Abstract (James), 13, 51
biomechanics, 132
birthday problem, 22–23
blackbody radiations, 53

Black Flag, 107
board games, 24–26
Boethius, Anicius Manlius Severinus, 25
Boethius's arithmetic, 25
Bohr, Harald, 132
Bonaventure, Saint (Giovanni di Fidanza), 86
Bose, Satyendra Nath, 23
Boulez, Pierre, 37
Breakthrough, 24
Brimberg, Jack, 72
Brix scale, 44
Buckland, Jonny, 106
Buffet, Jimmy, 106
Buildings, Antennae, Spans, and Earth (BASE) jumping, 48
Burgeson, John, 51
business, economics, and marketing
    stocks/stock market, 57
Butoku-kai, 80

cabbage, goat, wolf problem, 83
Caesar, Julius, 74
calculus and calculus education
    history of, 26
Calculus and Mathematica (C&M), 30
Calculus, Concepts, Computers and Cooperative Learning (C4L), 30
Calculus Consortium at Harvard (CCH), 30
Calculus for a New Century, 29
calculus in society, 26–31
    applications of, 28
    mathematics curriculum reform, 29
    traditional versus reformed calculus, 29
calendars
    number "12" and, 99
Campbell, Jack "Johnny", 32
canons, 36
Cantor, Georg, 87
card and dice magic, 76
Cardano, Gerolamo
    puzzles and games, 110
Cardano, Girolamo
    dice games and, 47
careers
    employers of mathematicians, 7
    in architecture, 62
    using algebra in, 3, 6, 7
Carey, Mariah, 106

Carroll, Lewis (Charles Dodgson)
    Alice in Wonderland, 21
    The Game of Logic, 25
    Through the Looking Glass, 1
    word puzzles and, 110
Casazza, Peter, 34
casinos, 19
Cézanne, Paul, 103
chai (life), 100
Chain-Links (Karinthy), 129
cheerleading, 31–33
Cherry, Colin, 34
chess, 1, 24
Chinese mathematics
    religion and, 88
Christian religious tenets, 85, 86, 98, 99
circles, 10
classic mathematical problems
    birthday problem, 22
    cabbage, goat, wolf problem, 83
    cocktail party problem, 34
    dining philosophers problem, 35
    Get Off the Earth puzzle, 110
    parallel climbers puzzle, 34
    party problems, 34, 35
    stable marriage problem, 80
    Tower of Hanoi puzzle, 82
    two-container problem, 82
Class Phenotype Probability, 23
Clay Mathematics Institute, 21
climbing, 33–34
cocktail party problem, 34–35
Codex Atlanticus (Leonardo da Vinci), 131
coding and encryption
    bible codes, 100
Coldplay, 106
Collins, Brent, 128
Columbus Dispatch (newspaper), 8
combinatorial game theory, 25
composing, 35–37
computational geometry, 59
computer aided design (CAD), 59, 144
computer aided looms (CAL), 144
computer numerical control (CNC), 60
Computer Speaks, The (Khalifa), 100
Connections in Geometry and Physics Conference, 39
connections in society, 37–42
    interconnected curriculum, 37
    mathematics as a universal language, 39

    nutrition labeling and mathematics, 39
    sports and mathematics, 38, 40
Conway, John H., 25
cooking, 42–45
Coonce, Harry B., 90
Coriolis effect, 21
Coriolis, Gaspard-Gustage de, 21
Critique of Practical Reason (Kant), 87
Critique of Pure Reason (Kant), 87
crochet and knitting, 45–46
cryptarithms, 110
cubists, 103
curriculum, K-12
    measurement systems, 96

Dai Fujiwara, 61
dancing
    ballet, 10
    ballroom, 11
    step and tap, 136
Debussy, Claude, 126
Dechales, Claude, 52
de Finetti, Bruno, 20
deism, 86
DeSalvo, Stephen, 139
descriptive geometry, 58
de Veaux, Richard, 140
De Viribus Quantitatis (Pacioli), 111
diatonic scales, 126
dice games, 46–48
Dietz, Ada
    Algebraic Expressions in Handwoven Textiles, 143
differential calculus, 26, 28
differential geometry, 59
dining philosophers problem, 35
Dionysius
    Parthenopaeus, 1
discrete geometry, 57
Dissertation Abstracts International, 90
distribution-free tests, 118
divination, 84
DNA
    molecular structure of, 73
Donald in Mathmagic Land (cartoon), 22
Dots & Boxes, 25
Dr. Atomic (Adams), 98
Drosnin, Michael, 100
drums, 105
Dudeney, Henry, 111

Index

Edidin, Dan, 34
Edison, Thomas, 32
Eight Ball, 21
Ein Sof, 87
Einstein, Albert
    Einstein on the Beach (Glass), 97
    theory of relativity and, 5
Einstein on the Beach (Glass), 97
Elements (Euclid)
    metaphysical reasoning and, 86
"E=MC2" album (Carey), 106
en l'air (ballet), 10
equal tuning, 114
Erdös number, 129
Erlangen Program, 141
Escher, M.C.
    symmetry and, 102, 141
Eskin, Alex, 21
Euclidean geometry, 58
Euclid of Alexandria
    golden ratio and, 65
    musical pitches and, 63
    Theory of Intervals, 63
Euler, Leonhard
    Seven Bridges of Königsburg and, 111
extreme sports, 48, 48–50

Fantasy Football Index (magazine and Web site), 51
fantasy sports leagues, 50
fashion design, 61
Fechner, Gustav, 66
Ferguson, Claire, 127
Ferguson, Helaman, 127
Fermat, Pierre de
    Blaise Pascal and, 19
    dice games and, 47
    Fermat's Last Tango and, 96
    probability and, 19
Fermat's Last Tango (Lessner & Rosenblaum), 96
Fibonacci, Leonardo
    Liber Abaci, 109
"Finite Simple Group (of Order Two)" (Klein Four Group), 107
fireworks, 52, 52–54
First, Outside, Inside, Last (FOIL), 8
Fischer, Gwen, 116
fishing, 54–55
football, 55–57
Fourier transforms, 34

Frézier, Amédée-François, 52
Fulke, William, 25
Fundamental Theorem of Calculus, 27

Gale, David, 25, 80
Galileo Galilei (Glass), 97
Galileo (Galileo Galilei)
    Star Messengers (Zimet and Maddow), 98
Game of Logic, The (Carroll), 25
game theory
    in baseball, 14, 51
    in basketball, 17
    in football, 57
Gardner, Martin, 78, 83, 113
    Mathematics, Magic and Mystery, 78
Garibaldi, Skip, 33
Gauss, Carl F.
    contributions of, 38
Gawrych, Billy, 50
geographic information systems (GIS), 59
geometric magic, 76
geometry and geometry education
    computational, 59
    dance as, 11, 12
    differential, 59
    discrete, 59
    early history of, 58
geometry in society, 57–62, 60
    design/manufacturing and, 59, 60
    early history of, 58
    fashion design and, 61
    graphics/visualization and, 60
    information systems and, 60
    occupational connections and, 62
    types of, 59
geometry of music, 63–65
Get Off the Earth puzzle, 110
Gilbert, W.S.
    Pirates of Penzance, 97
Gill, John, 33
Giotto di Bodone, 102
Girls' Guide to Fantasy Football (Web site), 52
Glass, Philip
    Einstein on the Beach, 97
    Galileo Galilei, 97
Gödel, Kurt
    Incompleteness theorem, 86
Goldberg Extension, 23
Golden Ratio, 65–67, 98, 103

Golden Rectangles, 98, 103
golden spirals, 66, 67
Goldin, Gerald, 120
Gottman, John, 108
Grand Design, The (Hawking), 86
graphs, 139
graph theory, 139
Greek gods and goddesses, 98
Greek mathematics
    Golden Ratio and, 65
    measurement and, 66
Green Card Lottery program, 74
group theory, 141
gymnastics, 67–69

Halma, 24
Hamiltonean Graph, 25
Hamilton, William Rowan, 25
Hanson, Howard, 37
harmonics, 35, 63, 65, 69–71, 147
Hart, George, 128
Hatori, Koshiro, 101
Hawking, Stephen, 86
Hawk, Tony, 49
Hein, Piet, 25
Hennessey, Andrew, 144
Heraclides, 1
Herzberg, Agnes, 139
Hex, 24, 25
HEXI, 35
Hickman, C. N., 10
Higson, Nigel, 127
Hildebrand, Harold "Dr. Andy", 107
Hines, Gregory, 136
hockey, 71–72
hockey stick graph, 72
Holmegaard bows, 9
homological algebra, 73
horse racing, 134, 135
Hubbard, Mont, 50
Huizinga, Johan, 24
Hull, Robert "Bobby", 71
Hurley, William, 72
Huzita-Hatori axioms, 101
Huzita, Humiaki, 101
hydrometers, 44
Hyperbolic Crochet Coral Reef, 128

imaging technologies, 94
Incompleteness theorem, 86
Indian mathematics
    religion and, 88, 98

Industrial Revolution
  mass production and, 144
infinity
  calculus and, 26
  religious tenets and, 85
information systems, 60
integral calculus, 26, 27, 28
International Baccalaureate Calculus (IB Calculus), 30
International System of Units (SI), 91
Islamic religious traditions, 99
iTunes, 107
Iwasawa theory, 97

Jacquard, Joseph, 143
Jacquard loom, 143
James, George William "Bill"
  Bill James Baseball Abstract, 13, 51
Japanese Railway Ministry, 80
Jeans, James, 115
Jesus, 99
Jewish religious tenets, 86, 100
jigsaw puzzles, 112
jiu jitsu, 80
Jonas, David, 104

kabala, 100
Kadison, Richard, 34
Kandinsky, Wassily, 104
Kant, Immanuel, 87
  Critique of Practical Reason, 87
  Critique of Pure Reason, 87
Kaput, James, 120, 121
Karinthy, Frigyes
  Chain-Links, 129
Kedlaya, Kiran, 2
Kelly criterion, 20
Kelly, Larry, Jr., 20
Kepler, Johannes
  laws of planetary motion, 7
Kepler's Laws, 7
Khalifa, Rashad
  The Computer Speaks, 100
Khovanov, Mikhail, 73
Kim, Scott, 113
Klein, Felix
  geometric spaces and, 59
Klein Four Group
  "Finite Simple Group (of Order Two)", 107
knots, 73–74
Kolmogorov, A. N., 20

Konane, 25

Labanotation, 11
Laban, Rudolf, 136
laws of planetary motion, 7
Le Balet Comique de la Reine (Beaujoyeulx), 11
Legendre, Adrien-Marie, 58, 141
Lehrer, Thomas, 106
Leibniz, Gottfried Wilhelm
  biographical information, 27
  plagiarism accusations, 27
Lenormand, Louis-Sébastien, 131
Leonardo da Vinci
  Codex Atlanticus, 131
  golden ratio and, 66
  paintings, 103
  sculptures, 127
  Vitruvian Man, 33
Leo XIII, Pope, 87
Lesser, Lawrence, 107
Lesser, Mary, 102
Lessner, Sydney
  Fermat's Last Tango and, 96
Liber Abaci (Fibonacci), 109
links, 73
"Lo Shu" magic square, 77
lotteries, 19, 74–75
Loyd, Samuel, 110
Lucas, Édouard (N.Claus), 82, 109
Ludus Algebraicus, 25
Ludus Astronomorum, 25
Ludus Latrunculorum, 25

Maclaurin, Colin, 5
Maddow, Ellen
  Star Messengers, 98
magic, 75–78
magic squares, cubes, and circles, 77
Mancala, 24
manufacturing design, 59, 60
Markov chains, 55
Markov decision process model, 106
marriage, 78–80
martial arts, 80–81
M A S S I V E, 107
mathcore, 36, 106
Mathematical Association of America (MAA), 29
Mathematical Games (column), 83
mathematical modeling
  archery bows, 10

  for climbing, 33
  hockey and, 71
  for swimming, 139
mathematical puzzles, 82–84
Mathematica (software), 30
mathematics and religion, 84–89
  Chinese religious tenets, 88
  Christian religious tenets, 85, 86
  divination, 84
  Indian religious tenets, 88
  Islamic religious tenets, 88
  pattern drawing, 84
Mathematics Genealogy Project, 89–91
Mathematics Genealogy Project Website, 90
Mathematics, Magic and Mystery (Gardner), 78
Mathematics of Marriage, The (Gottman, Murray, et al.), 79
Mathematics (record label), 106
math rock, 36, 106
Math Standards (NCTM), 29
McCready, Mike, 107
measurement in society, 91–96
  accuracy and precision, 93
  everyday applications, 93, 94
  Pre-K-12 curricula and, 95
  systems, 91
measurement, systems of, 91
Mega Millions, 74
Messiaen, Olivier, 126
metalcore, 36
metaphysics, 86
Metastasis (Xenakis), 37
Metromachia, 25
Micolich, Adam, 104
Milgram, Stanley, 129
millimeter wave scanners, 94
Mises, Richard von, 22
modes, musical, 126
Mondrian, Piet, 103
Moore, Eliakim, 38
Moss, Jamal, 106
Mozart, 66
Murray, H. J. R., 24
Murty, M. Ram, 139
music, 106, 107, 113, 114, 126, 138
  golden ratio in, 66
musical theater, 96–98
Music Intelligence Solutions, 107
Music IP, 107

Nabokov, Vladimir, 1
Nash, John, 25
Nation at Risk, A (NCEE), 29
National Basketball Association (NBA)
    Draft Lottery, 74
National Commission on Excellence in
    Education, 29
National Council of Teachers of Mathematics (NCTM)
    algebra education and, 3, 7
    interconnected curriculum and, 37
    reform recommendations, 29
National Football League, 55
National Hockey League (NHL), 71
National Research Council (NRC), 29
National Science Foundation, 37
National Security Agency (NSA), 6
navigation systems, 62
Nechunya ben Hakanah, 100
New Testament, 100
Newton, Paul, 139
Newton, Sir Isaac
    biographical information, 27
    Kepler's third law and, 7
    laws of motion/laws of gravity and, 81
Newton's laws, 7
Nicholas of Cusa, 86
Nim, 25
Nine Men Morris, 24, 145
Nomos Alpha (Xenakis), 106
nonparametric tests, 118
numbers and God, 98–101
    "7", 99
    "12", 99
    "19", 99
    bible codes, 100
    Golden Ratio, 98
    Pythagoras maxim, 98
    resurrection of jesus, 100
nutrition labeling, 39

Ockeghem, Johannes, 36
Ocneanu, Adrian, 127
"Octatube" (Oceanu), 127
Official Guide to Japan (Japanese Railway Ministry), 80
Okrent, Daniel, 51
Oliver, Dean, 15
Olympic Games, 12, 48
operations research (OR), 72
orbifolds, 65

origami, 101–102
origami technology, 101
Othello, 24
Ouranomachia, 25
Ouspensky, Peter
    Tertium Organum, 88
overblowing, 147
overtone series, 69

Pacioli, Luca
    De Viribus Quantitatis, 111
painting, 102–104
parallel climbers puzzle, 34
Parlett, David, 24
Parthenon, 98
Parthenopaeus (Dionysius), 1
party problems, 22, 34, 35
Pascal, Blaise
    dice games and, 47
    Pierre de Fermat and, 19
patterns
    drawings, 84
    figure skating, 130
    geometric, 17
    step and tap dancing, 137
Penrose tilings, 127
Pentominoes, 24
percussion instruments, 104–106
permutations and combinations, 22
Petteia, 25
philosophers
    dining philosophers problem, 35
Philosopher's Game (Rithmomachia), 25
Picasso, Pablo, 103
Pickover, Clifford, 100
Pirates of Penzance (Gilbert & Sullivan), 97
Pithoprakta (Xenakis), 37
Place of Mathematics in Modern Education, The (NCTM), 7
Platinum Blue, 107
Poincaré, Jules Henri, 73
Pollock, Jackson, 104
popular music, 106–108
Powerball, 74, 75
predicting divorce, 108–109
Principles and Standards for School
    Mathematics (NCTM), 29, 122
probability
    in baseball, 14
    in basketball, 15
    betting and, 19, 20

    birthday problem, 22, 23
    subjective, 20
probability theory, 1
Process Standards (NCTM), 29
Project 8 (video game), 49
Proofs from THE BOOK (Ziegler and Aigner), 86
PROVERB computer program, 2
puzzles, 109–113
pyramids, 66
Pythagoras of Samos, 5, 98, 111, 113, 114
Pythagorean and Fibonacci tuning, 113–115
Pythagorean numbers, 25
Pythagorean theorem
    algebra and, 5

quadratic equations, 7
quantum groups, 73
quarterback ratings, 55
queuing theory, 35
quilting, 115–117
Qur'an, 99

racquet games, 117–118
Rameau, Jean-Philippe
    Treatise on Harmony, 35
Ramsey, Frank, 35
Ramsey theory, 35
RAND Corporation, 3, 8
rankings, 118–119
Reagan, Ronald, 29
Realm of Algebra, The (Asimov), 3
reasoning and proof in society
    abstractions/symbolism and, 120
recreational mathematics, 83
Reeve, W. D., 7
Reidemeister, Kurt, 73
Reidemeister moves, 73
representations in society, 119–125
    in 21st century, 122
    internal/external structures, 120
    mathematics as language and, 124
    multiple approaches to, 124
    problem solving and, 123, 124
    translational skills and, 121, 122
representation theory, 120
resurrection of Jesus, 100
Rey, José-Manuel, 79
Reynolds, Simon, 106
Rhind papyrus
    puzzles and, 109

158 Index

Richman, Hal, 51
Riemann, Bernhard
    contributions of, 33
    geometric formulations and, 58
Riemann, Hugo, 65
Riemann hypothesis, 33
Rithmomachia (Philosopher's Game), 25
Robertson, Malcolm S., 90
Robinson, Michael, 17
Rogers, Everett, 51
rond de jambe à terre (ballet), 10
Rosenblaum, Joshua
    Fermat's Last Tango and, 96
Rotisserie League Baseball, 51
Royal Society of London, 27
Rubik's Cube, 112

sabermetrics, 13, 15
scales, 113, 125–127
Scènes de Ballet (Ashton), 11
Schneider, Robert, 36
Schoenberg, Arnold, 37
Schofield, Keith, 107
Scholastic Aptitude Test (SAT or SAT-M), 119
Scholasticism, 86
Schultheis, Michael, 102, 104
Scientific American (magazine), 83
sculpture, 127–128
security body scanners, 94
Sefer Yetzira (Book of Creation), 100
Séquin, Carlo, 128
Séquin-Collins Sculpture Generator, 128
Sesostris, King, 58
Seurat, Georges, 103
Seven Bridges of Königsburg, 111
shells and mortars, 53
signal processing, 34
Singer, Isadore "Iz", 34
SI units, 92
Six Degrees of Kevin Bacon, 128–130
SixDegrees.org, 130
skating, figure, 130–131
skydiving, 131–132
Slint
    Spiderland, 106
"small world" phenomenon, 128
soccer, 132–134
Society for American Baseball Research (SABR), 13
Sophocles, 1
speedometers, 41

Spiderland (Slint), 106
sport handicapping, 134–135
sports, 13, 15, 38, 40, 41
sports dngineering/equipment, 50
square dancing, 135–136
squaring a double-digit number, 83
stable marriage problem, 80
Stable Matching Algorithm, 80
Star Messengers (Zimet and Maddow), 98
statistics
    sports, 13, 15, 40, 41
step and tap dancing, 136–137
Stockhausen, Karlheinz, 37
stocks/stock market, 57
Strat-O-Matic, 51
Stravinsky, Igor, 11
string instruments, 137–138
sudoku, 138–139
Sullivan, Arthur
    Pirates of Penzance, 97
Swanson, Irena, 116
swimming, 139–140
Swinburne, Richard, 100
Sylvester, James Joseph, 62
symbolism
    representations and, 119, 124
symmetry, 63, 65, 140–143
Système International d'Unitès (SI), 92

Taimina, Daina, 45, 128
Taniyama-Shimura Conjecture
    Fermat's Last Tango and, 96
Tartaglia, Niccolo, 109
tartan setts, 144
Taylor, Richard, 104
tempered scale, 63
Tertium Organum (Ouspensky), 88
tessellations
    in applied design, 116, 141
    in Islamic architecture, 89
textile industry, 144
textiles, 143–144
"That's Mathematics" (Lehrer), 106
Theory of Intervals (Euclid), 63
theory of relativity, 5, 59
Those Parts of Geometry Needed by Craftsmen (Abu'l Wafa), 89
Through the Looking Glass (Carroll), 1
Thurston, William, 61
Tic-Tac-Toe, 144–145
tilings, 116

tonal harmonies, 63, 65
Tonnetz (Tonal Network), 64, 65
topoisomerases, 73
topological puzzles, 111
topological quantum field theory, 73
topology, 73
tournaments, 2
Tower of Hanoi puzzle, 82
traditional versus reformed calculus, 29, 31
trampolining, 68
transformations
    compositional, 36, 37
Transportation Security Administration (TSA), 94
Traveller's Dodecahedron, 25
Travers, Jeffrey, 129
Treatise on Harmony (Rameau), 35
Tremain, Janet, 34
tube resonators, 147
Tucker, Alan, 34
Turner, Joseph, 104
twining, 18
Twixt, 24
two-container problem, 82

Ulrish, Richard, 128
USA Mathematical Olympiad, 3
USA Today (newspaper), 41
U.S. Bureau of Labor Statistics, 7
U.S. Military Academy, 58
US News and World Report (journal), 118

van Gogh, Vincent, 104
Vassiliev, Victor, 73
Vedic mathematics, 88
visual symmetry, 141
Vitruvian Man (Leonardo da Vinci), 33
vocal cords, 147
voice, human, 147
volleyball, 145–147
Voronoi diagrams, 60

Wainer, H., 140
Walden, Byron, 2
Wall Street Journal (newspaper), 98
Washington, George, 74
weaving, 143
Webern, Anton, 37
Weyl, Hermann, 6
Whitehead, Alfred North, 7

White, Shaun, 50
whole-tone scales, 126
Wiles, Andrew
    Fermat's Last Tango (Rosenblum), 96, 97
wind and wind power, 147
wind instruments, 147–148
Witten, Edward, 73
World Association of Veteran Athletes (WAVA), 135
World Cup Finals, 38
World Squash Federation, 117
Wynne, Arthur, 2

Xenakis, Iannis
    Metastasis, 37
    Nomos Alpha, 106
    Pithoprakta, 37
X Games, 48
X & Y album (Coldplay), 106

Yackel, Carolyn, 45
Yang-Baxter equations, 73

Ziegler, Günter
    Proofs from The Book, 86
Zimet, Paul
    Star Messengers, 98
zodiac, 99